UNIVERSAL

DIRECTION

BOOK 1

BY ANDREW TRAPPES

First Edition

Cover artwork by www.ebooklaunch.com

Typset in Gentium Book Basic

Author enquiries to: andrew.trappes@gmail.com

Contents

For my wife

Chapter 1

The Cataclysm

THE MEDENS FLOATED STOICALLY IN space, its shape like that of a giant manta ray, unmoving as if sleeping in a currentless eddy. A ship of the Organisation Universal Fleet, its open mouth provided an indiscernible indicator of its lonely path amongst the stars. A smaller vessel, much like a boat drifting away from the shore, pushed off from the *Medens* and ebbed away.

With an exhale of its back engines the smaller craft vanished from view. The OUF *Medens* now sat alone, the recent departure serving to highlight the stillness of its position in the stars. A short burst from its own engines confirmed life to the solitary ship and its occupants. Though slow in motion, the *Medens* was back on its path, healing the roads of space broken many years ago.

Hames felt his knee buckle and wither. A sharp twang in his left leg rang throughout his body, telling him an expansive story in the first breaths of the life of a second. He fell to the floor of the hard grey plastic gangway and knew immediately the cruciate had gone again.

He grabbed the med-bar from his small square satchel and wiped a black smudge from its screen. Through the pain and discomfort he breathed rhythmically, and carefully programmed the device based on what he believed, fairly confidently, had happened to his knee.

While Hames was over sixty years old he looked at least a decade younger; his youthful skin and eyes belied a core feeling of age and dilapidation. This most recent event served to confirm his progression. Years spent amongst the stars had protected him from most terrestrial injuries. However, to recent notice, a new physical enemy had presented itself – the degradation of time that accompanied his long adventures across the Universe.

Sitting there on the elongated mesh gangway, Hames waited for the med-bar to charge. Alone and incapacitated, he knew it wouldn't be long until his mind raced towards her and the never-ending possibilities of what may have happened. *Don't look at them, don't look at your hands.*

Gazing upwards, he focused on the strip of lights placed at head height along the plain grey corridor that provided neutral illumination to all those walking on the deep-space vessel. Then below him, underneath and enclosed by the gangway, were the cords and compartments that ran without end, pushed to the side by the wide tube-like conduit that glowed blue and bright. Surging, powerful energy gushed and hummed, its two-step tone producing a binaural beat that bounced off the flat grey walls of the *Medens'* walkways.

This was the background noise to his existence; the constant hum of the ship's arteries, giving energy and life to all aboard. A pause in the distraction drew him to the inevitable; his eyes searched towards to his hands.

On the back of his left hand Hames saw the faded tattoo of a compass, outlined in a shade of red. The circle picture with two pointed triangles coming off two sides was atypical. Black and cream shades filled the middle with arrows pointing north and south. On his opposite hand, Hames saw another tattoo; a similar compass retaining the soft red, creams and black indentations. This compass, though, was broken, split down the middle in a jagged line, as if an earthquake had run through it.

The two compasses were always present in everything he did although their origins were only sporadically acknowledged. Except for touching the unbroken compass that held his communicator, weeks would pass by without a deep consideration of the purpose of these pictures. He sometimes wondered why he never had them removed. However, this was a recurring thought and the answer was always the same.

The med-bar charge came online. As intended, it would speed up the blood, nutrients and effort that in the natural course of events would be assigned to this injury. Hames felt trepidation yet reasoned this away with the realisation that nutrients are by any natural perception just another form of energy.

Sighing, he gripped his upper right thigh, the effort expanding the fracture in the broken compass on his right hand wider and wider. He directed the device to the problem area and paused, listening to the isochronic beat of the energy conduit below. Once engaged, the med-bar pushed light and power in a stinging shot throughout his knee. Then it was done. The standing advice for use of all these activities was to rest for eight minutes while tendon and muscular repair actions settled. So here he was, stuck, lying down on the connecting pathway leading to the repair control room.

Hames considered whether he would be found by another crew member; it was unlikely given the size of the ship. Sitting up, he felt his body lift slightly off the gangway, first his feet and then his legs. Without warning, he took a short drop back down onto the gangway, twinging his knee again. On the connecting lines between modules the artificial gravity was less reliable and Hames recognised he was at one of those points in the ship. He looked around and saw a sign above him:

OUF – Medens – Safety Missive

*In all connection passageways please walk slowly
and steadily. Do not run.*

*When moving through the module have one point
of hand contact with the rail at all times due to
gravitational fluctuations.*

The *Medens'* dependence on small spinning plates located in the lower sections of the ship meant its gravity well was dispersed across the operational and living areas. It was just unfortunate for him that in certain points the well was not completely filled. Despite his athletic arms and broad shoulders, he was slight of build, an average height. There was not a great amount of inherent weight for him to account for. Hence he should have paid more attention to the safety warning for travelling through the ship. Instead, his mind had strayed to the day ahead and the new addition to the Mission Team waiting for him in the repair control room.

Hames tried his knee again; finally he could move it. Breathing out, he braced his heart for the next effort, aligning it with the beating life of the *Medens'* energy flows, then raised himself up. Capability re-entered his body with every step as he eased his way along the gangway towards the repair control room. As he approached his destination, the white clean wall in front of him parted and morphed its structure to allow entry to a round room with a glass ceiling that showered him in stars through the view above.

All around the brightly lit space, walls of data and cartography welcomed him to their colourful arrays. This did nothing to impress Hames, as for all its effects the goal of the *Medens* was quite simple: to repair space. Space that had been damaged from the use of a technology that allowed travel across the Milky Way in an instant, but which ultimately broke the very pathways it voyaged across. This technology had come at a high price, to him and others around him.

He searched the room; he was still alone. The new officer joining his mission team had not arrived. This was surprising as, given his own delay on the gangway, he had expected her to beat him to the start of the day.

In the absence of the new officer, Hames began a series of checks to ensure all activities were meeting elected targets and performance. Out of curiosity rather than impatience, he pressed the communicator in his compass to hail Junior Mission Officer Emmalyne Biggs. There was no response. Looking up to the star map to his left, he saw his reflection in the sharp dark glass. His father's grey and blond flecked beard stared back at him. There was an organic feel to it and its colours seemed to balance his head. Mostly, the beard served to hide his chin, which, while a perfectly fine feature, had been rather prominent in his younger days and was a source of much comment. It was the thing he should have gotten past when he was older. However, Hames was old and letting go of past grievances wasn't in his skill set.

In his reflection his deep blue eyes saw a name patch that read 'Hames Naughton, Chief Mission Officer'. This told those who read it that he was tasked with maintaining the equipment on board their ship, the OUF *Medens*. Hames enjoyed being the 'Chief' and being in charge of the principal reason for their mission, conducted so far out here in deep space. He served as the last point of call for key decisions, not that there were many of these, and the weight of responsibility helped pass the time on the long journey.

One of these responsibilities was the training of new mission officers on board the *Medens*; a job that was becoming more and more infrequent the further away from Earth they travelled. The most recent recruit to the

Medens came four days ago and like other long-term travellers had spent an extended time in cryo-stasis hibernation – nearly two years. The first few days of her time on board would have been a constant internal battle urging her body to wake up.

On the screen closest to him Hames called up Emmalyne Biggs' profile. Like anyone selected for his group, her data showed her to have a high mental capacity, training in outer system computations and the expected experience in particle repair technology. Multiple certifications showed a comfort in deep-space life and all the rigours that this entailed. Her core aptitude scores were unparalleled although she was chosen for this trip for many reasons other than her extreme intelligence. She was willing to come and make the sacrifices required to be on the *Medens*.

Notably, Emmalyne was youthful, although past adolescence. Hames had been compelled to look to both the young and the old by the Organisation before filling his Mission Team; these distinct generations being more suited to long-term travel. Given she was in the former category, Hames would need to complete her training on board and test her understanding in practice. Emmalyne would need to be ready for the most important event that the *Medens* had faced in its three-year journey. In five weeks they would make contact with the very first outer solar system colonised by humans nearly two hundred years ago.

While he waited, Hames programmed the computer for an update:

> "In the last 12 hours the Medens has travelled 112,345 kilometres and to date has repaired 0.00392% of the Mermai Tentacle."

Since docking with the resupply ship, the ship had been travelling at a low rate. This meant the *Medens* had repaired next to nothing in the last few days. Despite the incremental progress of their mission, Hames knew it was important to check their procedures as the technology used to repair space was volatile and expansive.

Finally, Emmalyne responded on the communicator. "I'm coming; there in a minute."

Hames looked up at the stars, wondering what she would be like. He felt a slight excitement in his mind at the prospect of meeting someone new, an infrequent occurrence on the *Medens*. His examination of the constellations was disturbed by the negative sound of the entry point to the repair control room refusing to part. Again the noise pinged to suggest someone had been denied entry. Finally, through the room communicator came a voice: "It's Junior Mission Officer Emmalyne Biggs here. I'm having this problem across the ship – it's not recognising me!"

Hames tapped the screen in front of him and the image of the aspiring entrant appeared. "Hi, it's the Chief here. I'll have the entry point open in a second." The ship's log revealed the problem as a simple one: Emmalyne's bio signature had only been given access to visitor areas. After he rectified this the white clean wall finally spread open.

Emmalyne entered the room at speed, her eyes checking every aspect of the cornerless room. It was apparent to Hames that she was not entirely calm. Her high cheekbones gave her face a sculptured look and a slight frown on her forehead appeared to elevate them even further. This made her look more like a disapproving statue from a Service Park institution than a junior mission officer. That her blonde hair was pulled back tight in a ponytail, and that she was quite tall, did not detract from this vision.

"Hello. Chief Hames Naughton." An extended hand reached for Emmalyne, who recoiled at Hames' broken compass tattoo.

After this pause, his new teammate returned the courtesy of shaking his hand, although this pleasantry was not reflected in her voice. "What is wrong with your hand? Did you burn it somehow?"

Hames held up his hand to provide a longer viewing and smiled. "It's a skin painting."

"What's it for?"

Hames' smile turned to laughter. "It's a picture to remind me of something important. People used to have them all the time."

Emmalyne looked more disgusted than curious and did nothing to hide this in her voice. "Really?"

"Really."

"Does it hurt? It still looks burnt to me with that large red crack through it." The new officer was wiping clean her hand in mid-air.

"No, it's fine, thank you. The left-hand image holds and hides my communicator, so it's kind of useful."

The two officers moved away from the entry point, allowing the space behind to morph into the form of a wall, concealing them from the long grey corridor.

Emmalyne paced the room looking up and down at the screens, her defined face scrutinising all she could see. "I'm sorry I was late."

"It's okay. If you'd been here on time I could've met you at the door, saved you some difficulty."

"I know, I'm sorry." Emmalyne grabbed her forehead. "I'm still a little fuzzy, still having problems focusing on time by the minute." The rubbing now focused on Emmalyne's deep-hazel eyes.

"Didn't your communicator go off? It should be linked in with the schedule of duties for all officers on board the *Medens*." Hames was polite and quiet with his question.

"I didn't hear." Emmalyne looked confused at herself for the missing of such a small and simple task. The room went quiet, only the distant hum of the energy conduits in the corridors outside broke the silence. The moment felt peaceful and allowed Hames to analyse his new co-worker. She appeared to be rushed and distracted, probably suffering from the effects of the stasis-hibernation.

Emmalyne ran her hand through her blonde ponytail. Hames noticed the familiar brown strap around her wrist with a golden symbol. Clinging tightly to her thin arm it looked a part of her, the edges between skin and material imperceptible. The golden shape was well known to him; the eight circles intertwined with each other to produce the Flower of Connection, the symbol for Universal Direction. To Hames it had always looked more like an eight circle cyclonic vortex, an old argument for him which he did not seek to revisit. The connected circle image periodically spun and glowed showing an active life of the participant who wore it.

"Okay, let's start. You've been in these types of control rooms before, do you see anything familiar?"

"I see that there are lots of maps, and computers and targets that we don't seem to be achieving. So it's about what I expected for repair space travel." Emmalyne looked unbothered by giving this assessment.

"It's true; we don't seem to be making great time at this, for a range of reasons. Let's start at the beginning and see if you can tell me why our progress is so slow." Hames presented this question to her with one hand open-palmed.

"What do you mean the beginning; the start of the tentacle from Dover to here?" asked Emmalyne.

Hames was aware his next request may be challening; however, he needed to get a grasp of her knowledge and see how she was feeling overall. "No, the real beginning. The start of this whole journey. Why are we repairing these fractures in space?"

"Oh come on, this is a waste of time, we both know what happened. Is there a joke somewhere in this?" Emmalyne protested.

"Not at all. I'd just like you to show me your understanding of the broader predicament the *Medens* is addressing and in doing so you'll hopefully be on the right path for understanding my ways. Which is kind of important as I'm the Chief around here." Hames grinned in a way that he hoped to convey she had no choice in the matter and had to comply with his request. "Tell me, what is your objective observation of the history that brought us here?"

"What does that matter?"

"I guess I want to test how you're thinking, see how affected you are by the experience in stasis."

"I know some of my thoughts scattered." Emmalyne paused to swallow in her moment of introspection. "That doesn't matter though. I'm still awake and experiencing everything perfectly fine."

"It sounds like in your first few days you've been a little lost."

Emmlayne challenged him with an upwards twist of her head. "Isn't that normal?"

"I guess so. Using a common point, though, like history, is a useful way to see if you're objectively understanding the world around you."

"History isn't objective, it's shaped by how we view it. The qualia I feel and experience are there for me alone. How I experience the world is from my point of view only. What I see and think when I walk into this room is different from you. My qualia are mine. We can compare and correlate but it is all subjective."

This line of discussion raised some concerns in Chief Hames. He realised then that the leather bracelet with the Universal Direction symbol was not just for show with Emmalyne. She was committed. Nonetheless, he wanted a plain recounting of recent Earth history, to test her memory and recall of well-known events.

"Maybe everything is subjective. But a chair is a chair; a square has four right angles; there's always a certain sequence to time and events. I understand the subjective nature of the past; however, let's find a common point by looking back at the Cataclysm."

Emmalyne relaxed her back against a workbench and returned a tight smile. Her voice was slightly more welcoming than before. "You don't need to treat me like a child. I'm not feeling perfect. I'll come right soon."

Emmalyne was clearly not a child; she was considerably taller than he, and the hint of fragility in her voice belied a strong resolve. Hames raised his thick arms to grab at a screen-map displaying a long route through stars. He pulled down one of the route sections to the screen in front of him. "Indulge me," he requested as he began to work away on the data ahead of him.

"Okay, I'll play along." She agreed to the test, but Hames could see she was managing herself and possibly fighting a need to tell him to have a long cold sleep. For now, though, she appeared to have herself under control. Emmalyne commenced. "About forty years ago, um, no, let's go further back, over a hundred years ago a group of scientists based in orbit around the moon were working on transforming the basic molecules around us. By accident they stumbled onto what was to become known as 'particle conversion drive' or 'PCD travel'. It was 2305 when they identified the pushing and breaking of particles through a molecule separation process could produce a motion of such magnitude it looked like objects disappeared."

Hames cut in. "Some people referred to it as folding space; it's not really that, although it's close to it."

Emmalyne continued as if he had never spoken. "And soon they realised the potential for this to be a form of space travel unlike we'd ever seen before. It was much, much faster than the light-speed travel already in place. It was a form of travel only limited by a need for a clean path ahead of it. You know, no planets, asteroids or anything – it had to be pure space."

"Do you remember the first probe from the Archon Project back in the 2132?" Hames turned to check his screen, encouraging her analysis with a gesture of his hand.

Emmalyne sighed. "I think so. That probe smashed into the Oort cloud. Telling us we had to clear a pathway out of our solar system."

Hames agreed with her. "Particle conversion, using it to step through space, any interruption can cause a fairly big reaction."

A small pause from the new officer left an uncomfortable silence in the room, her voice sounding tense when she finally responded, "Thanks, I think I know that."

There was something about Emmalyne that was both familiar and unseen by Hames for some time. Her words conveyed comfort yet her face told a story of something else; it was hard to pinpoint. Hames examined her again as she detailed the spread of colonies to the outer reaches of the Milky Way. He had asked her to conduct the simplest of recountings, one that a child of seven or eight years old could provide. Of course she was frustrated. According to her profile she held one of the strongest comprehensions of particle conversion technology ever recorded.

Hames gave her some encouragement. "You're doing well at this, just relax. Now tell me how the efforts for particle conversion drive travel achieved the clean lines through space."

"Oh, I guess it was eventually understood that a straight line of safe travel was hard to guarantee and pathways could only be plotted with incredibly powerful telescopes and super bio-computers based in Earth's orbit. These computers programmed and locked the pathways that the ships enabled with PCD could follow. In the end, there were only a few points that were deemed both safe and worthwhile to travel to. The ships weren't equipped with the intensive range of equipment required to plot a new and safe course even deeper into the Universe. They'd just be lost if they did."

Hames completed the next part of the tale. "After this, a whole load of migration occurred out to the set destinations; seven of the pathways were to viable worlds."

"Have you travelled by PCD?" Emmalyne halted her recounting, the suspicion in her voice suggesting she was now aware of the depth of Hames' involvement in this topic.

Hames looked up through his blue eyes and made a conscious decision to focus on his experience of interstellar travel and that alone. "Yes, yes I have. It was amazing. One moment you would be docked just outside of Earth then minutes later you would be in a completely different part of the Universe. What was even more amazing was how quickly we, as a community, adapted to it. How quickly we took to the stars and made new homes. In single generations, bodies began to adapt as well as cultures. Let's stay on track, though. I'm interested in your understanding of the basic reason the *Medens* is out here. So," Hames again half-raised his arm with an open palm, "please continue."

Emmalyne focused on Hames' gesture as she fidgeted with loose wisps of her blonde hair. "Okay, okay. After a hundred years of PCD travel the Cataclysm occurred." Emmalyne looked to Hames, trying to spot any noticeable change at this statement but he offered her none. "No-one knows why the particle drive tech blew up or ignited, just that it all started with a ship called the *Python*." This time Hames' neck flexed to show his discomfort, although Emmalyne didn't notice. "I've read all the theories about what and why it happened; one is probably right. One thing is certain: all the PCD pathways shattered, leaving an invisible trail of debris. An unrecognisable mess which meant that particle conversion travel could never be used again. Ships trying the pathways were obliterated thoroughly and seemed to cause more damage. It was a turning point for the human race. Devastating on many levels." As Emmalyne finished Hames felt his face drop, enough to trigger her perception. "Chief, are you okay?"

"I'm good," he reassured her. "Like most people around at that time I was affected by it all. Most people were." He tilted his head to one side in recognition. "The small understanding we have is that we were converting photons of light and dark matter to create energy. We didn't know what type of position we were establishing in the Universe. Like a collapse of wave motion, there was a collapse of light. Once the particles were ignited in the Cataclysm they blackened the pathways to our colonies."

Emmalyne reflected on his last comment before replying, "Eventually the Organisation came along and made a plan to fix the communities and space broken by the PCD Cataclysm."

"You make that part seem so quick. The technology to fix the broken space was just as much a breakthrough as developing the PCD tech in the first place."

Emmalyne interrupted him. "And this is where ships like the *Medens* come into action. It's our mission to fix the whole damn mess. The idea is that future generations might be able to travel again to the other colonies. To see the deep Universe and safely move about the stars without ripping them apart. The routes where ships used to travel out to the seven colonies are what we're repairing." The young mission officer stood up and moved next to her Chief and the large map of stars with a long unending line stretching out ahead, the path of the *Medens*. "When all the routes to the colonies are looked at on a map they look like an octopus stretching out its arms from Earth. That's why some people call the routes 'tentacles'." Emmalyne began to run her finger over the long line covering the cartography in front of her. "The *Medens* is repairing the tentacle that stretches out from Earth to Mermai."

"That's really good; you don't seem to have any memory issues. You've retold the history as it is." Hames turned to another screen and made an entry to confirm this while Emmalyne sighed and looked up skywards through the clear dome to the stars above. A quiet hum in the room provided a feeling of warmth as barely noticeable flashes and reports ticked over on the screens around them. Hames attempted to break the moment. "The *Medens*...it's a long journey ahead. It must have been a big decision to come out here."

Emmalyne turned back to him as her eyes stayed focused on the stars above. "We broke it you know. Space. I mean the collective we, not us personally. So strange that we could reach out from our tiny little planet and do such a thing. All my life it's been broken. To fix such a thing, it would be..." she paused and her hazel eyes became glassy. "It could have a great impact on people. It's a powerful reason to be here."

Hames quietly nodded. He didn't know why the pathways to far off planets had been destroyed. He had been there when it first happened. He had seen and felt its devastation. Perhaps one day he and this young girl could be there when the darkened roads could finally be renewed.

Chapter 2

Ugor Eels

HAMES OPENED HIS EYES AND closed them again. The picture in his mind was distorted: a view of deep-brown beaches stretching for miles, a green field behind and then, maybe a mountain, no, a hill. It had been a strangely real vision just before he awoke, but now it was fading. Replacing this past vivid picture was the clean grey lines of his small but efficient home-pod.

No more than two by two metres, the pod accommodated a bunk, a small desk at one end, and various storage solutions. He sat up in his sheet and considered the dream. *Was it a beach or a field? I think one of them was green.* The dream began to fade. He reached over to his right and pressed a button with multiple circles on it. Instantly his bedding retracted into the wall and a panel formed over its exit point. Hames alighted from the bunk, his feet barely making a sound on the warm plastic floor.

He slowly touched a space to the left of his bed; a screen formed on the wall and presented a mirror for his use. Looking back at his greying self he reached into his mind, but the dream was gone. During sleep he'd held a reasonable grasp of what had been occupying his perception but now it had completely vanished. Nothing more about the dream troubled him as he began to ready himself to meet Emmalyne at the exit point from the home-pods.

Standing at the desk next to his bunk, as he did most mornings, Hames surveyed the plastic notes and calculations suggesting that overnight a great effort of industry and innovation had been undertaken by an ingenious mind. The array of work, however, was just his. Working at night was a nocturnal ritual that at times felt unending and absent of outcome. Another evening wandering down dead hole after dead hole. In any case, the Universe was infinite and there would be more paths to investigate again tonight, and the night after that. He wondered if Emmalyne would be any help in his work; she was exceptionally bright.

It had been days since she had come aboard and the two of them had worked closely with each other. So far they had spent three shifts together, executing the tasks required within the dome-shaped repair control room. The incidents of erratic behaviour he had observed at their first meeting had persisted. There would be moments of peace and progression as they moved about their work. At other times her voice would quiver or raise. At some points even tears had flowed. Hames had seen this before: the effect of being in cryo-stasis for such a long time temporarily altered a person's body in ways he didn't fully understand. It seemed to be more prominent in Emmalyne.

Moving to a clothes cupboard opposite his bunk, Hames chose his blue mission uniform, pulling it over a white undershirt that clung tightly to his thick-set arms. At first he reached for the typical flat white slip-on shoes, but instead chose a pair of chunky brown lace-up boots. Finally he was ready for the short walk to the waiting area outside the entrance to the Home-Pod module. The area was empty except for a small padded bench against a bleak white wall. It was no surprise that Emmalyne wasn't there waiting for him. In truth, he was in no rush and he had patience for her situation.

After pacing up and down a nearby corridor, he stopped in front of a side circular window looking out to the night. The stars were still, as was the *Medens*. Despite this abeyance of activity, Hames was sure that time held its step for nobody. It was persistent and evidentiary in recording the incidents of the past. Decades ago he had witnessed the crisis following the Cataclysm when Universal Direction had grown in response to the pain of those around him. After that period, many years later, he had observed a general feeling of betterment in the management of internal challenges. The community had grown resilient as a whole. Nowadays, not many people, except those who had experienced extreme events such as being in stasis for years, spoke or acted out of order.

Hames was unprepared for Emmalyne's intense mood swings. One moment, a day ago, the rejection of a scanning request into the repair program had produced much emotion and tears. Reaching over in his stiff manner, Hames had patted her on her left shoulder, as if he was trying to loosen the last contents of a steel cereal container. This effort had resulted in her reaching to him and wrapping her arms around his wide shoulders. Her tears and flood of feeling had in turn prompted more robotic patting actions. There was no question about it; he was terrible at dealing with emotions and knowing how to respond to them.

Today as she walked towards him from the home-pods she seemed improved and somewhat lighter. They leant together on the grey bar that faced outwards to the darkness. Hames offered warmth and friendship in his voice without looking directly at his colleague. "Okay, we have the tentacle repair course mapped out in preparation of the move back into light alignment drive. That work's all done so we have a bit of space on our hands." Emmalyne shook her head at this tired joke from the chief mission officer. "So this morning I want to show you the bridge and you can meet the person responsible for everything that happens on the *Medens*, the chief operations officer or COO."

Before they started to leave their porthole view, Hames speculated that a pre-emptive counter to any onslaught of emotion may prove a wise idea. He stumblingly started with, "I, ah..." He then directed his focus to what he needed to actually talk to her about and said, "I wanted to check on how you are doing, okay? Are you, okay?"

"I guess so. I feel a bit better today." Emmalyne smiled with a shrug. "I'm okay most of the time and then something happens. I feel short of breath and like, like something bad is going to happen." Her face was no longer smiling. "I really don't remember feeling this way ever before."

"It's understandable and you expected you would feel some sort of difference coming out of stasis." Over the last two days Hames had told her similar things in different ways. He knew from his youth and the time after the Cataclysm that he should not make her feel wrong in any way for her imbalance. Like most people, he'd felt intensely imbalanced when he found himself trapped on Earth. Not knowing what had happened, the absence of clarity had torn at him deeply. It was a scar that endured to this day.

Emmalyne looked out to the night. "Chief, I'm pretty keen to see more of what life is like on the ship, meet the COO and see the change into light alignment mode, I think it will help, um, elevate me."

In the few days they had known each other his regard and sympathy for the smart if somewhat troubled young woman was high. "You're elevated already in my view."

Emmalyne put her arm around his shoulder. "Chief, you're great."

Feeling self-conscious, Hames changed the subject and began their path for the day. "Let's head to the bridge." The two crew members progressed through the corridors of the *Medens* on an ever widening gangway till they reached a hard grey wall. It appeared to be a dead end at the farthest left side of the *Medens*. The conduits beneath them stopped at this barrier and the quiet hum of power lines below echoed back at them, for a moment making Hames feel unsteady in the artificial gravity of their vessel. But on their arrival the wall shifted and moulded itself back to four evenly spaced sides, revealing a square-shaped entry point to the control centre for the *Medens*. The sparse design and sound of their previous path gave way to the activity in the small room ahead of them.

The square-shaped area was darkened; a few lights and screens were all that illuminated the limited structure within. Two pilot seats sat before a set of panels, each resembling the half shape of an octagon, both with diverse screens and controls. A young officer sat in one of the seats. Behind this arrangement to the right of the room was a similar seat and station, holding within it an older woman with grey overalls that were topped at the shoulder by a red patch that ran from her neckline to her heart. Standing up to greet them, long chocolate and grey-flecked hair fell around her deep olive-and green-skinned face.

Emmalyne was surprised by the person in front of her. "Oh, my mind, you're Mermain?"

"And your chief operations officer, pleased to meet you, Emmalyne. I'm Melini, Melini Guerra." A worn olive-skinned hand extended itself to the young mission officer.

Emmalyne greeted her superior, but with far too much caution for such an introduction. "My apologies, I haven't met anyone from Mermai before. I've only seen vid-clips of Mermai and its people."

"That's not a problem. It's happened more to me in the last ten years than I care to count. There's not many of us left from Mermai who were stranded after the Cataclysm. How's the Chief treating you?"

"Apart from the odd childlike question or two, not so bad." Emmalyne gave a small smile to Hames, who simply scrunched his beard in acknowledgement.

The COO then placed one hand on the back of her chair, ready to return to her work. "Well, it's been good to meet you. Any time you need to talk or have any questions please come see me."

"I do have one question. Why is it so dark in here?"

Melini Guerra looked surprised. "Okay, interesting question. It's not my preference but Leena here likes it that way." The COO motioned to the person in the pilot seat. Pressing the button on two screens to close them down, a small young woman with impeccably short brown hair stood up. "This is our first pilot, Leena. She's convinced me it's better to work in this blacked-out way." The COO scowled ruefully at the small first pilot and shook her head.

Hames was aware of this old and well-trodden argument between the leader of the *Medens* and her pilot. Leena could indeed be a firm opponent in a discussion; her prominent sharp nose seemed to carry her towards arguments with most people across the *Medens*. Her deep-set dark eyes and stocky features presented an image of resolution. However, these notable features were overshadowed by the golden circular shape that was embedded to the front right side of her neck.

Leena spoke confidently and with no intent to threaten as she reached over the back of the seat. "As Melini said, I'm Leena, first pilot."

Emmalyne took her hand and stared at the neck of her new crewmate. "The spot on your neck, what is it?" Emmalyne leant down and around as if trying to gauge how big it was. "It's almost as big as my palm."

Leena appeared to do her best to manage the discomfort from this question, only slightly curling her upper lip. "Have you not seen a Sentir before? You must have been in stasis for decades, not years." Absent-mindedly she began to rub the large golden shape then checked herself to a halt.

"Oh, I've seen them; they were the size of a pin. Is that what yours is? It's so big and bright! They were meant to be black I thought?"

Leena's frustration was beginning to boil over. "In case you haven't noticed, my skin is fairly dark; my father's from Goa." The first pilot gesticulated around her neck. "The *Medens* has to be able to locate my Sentir when I pilot the ship."

Emmalyne gave her a challenging look that bordered on disgust. "A Sentir? Really though? A physical connection to breach locality? Isn't that only for Savai fanatics who believe they need tools to make a connection to another entity."

"Those fanatics are not, in fact, fanatics," Leena spat back. "And I am one of those um, non-fanatics who study in the Savai branch of Universal Direction."

"Oh." Emmalyne looked away at the end of her sentence. Feigning ignorance she asked, "Wasn't the Savai branch shown to be unaligned with Universal Direction?"

"I'm as committed as the next person to Uni-D. I just think I get a better pilot experience out of the ship when I connect directly to its operating system."

"You're already connected though," Emmalyne shot back.

Leena frowned and placed her hands by her sides, trying a calming technique of sorts. "My consciousness is mine, it's exclusive."

"That is such an old statement, you sound almost as bad as the Chief here."

Hames shifted in his stance at the suggestion he was the low bar for the point of comparison in the beginnings of this debate.

Leena scrunched her pointed nose and hints of stress began to show in the brow below her short brown hair. "No, that is unfair. I see the subjective truth of Universal Direction just as much as anyone else."

The two women continued to examine each other in silence and Hames waited for potential hostilities to flare. He was grateful at that moment when a light tone rang out from a screen to the right of the COO's chair. Both Hames and Emmalyne directed their attention over the shoulder of Melini Guerra, curious as to what was happening.

"Oh, that's unexpected but good. I'm getting some readings that suggest something pretty special could be happening." Melini Guerra continued to check and review the data in front of her.

Hames and Emmalyne looked at each other, still processing what the COO had said.

"I'd better let the crew know." The COO pressed a nearby pad and spoke throughout the ship. "Hello all, we've picked up some curious readings on the bridge here. Something you may not believe. It looks like we could have some Ugor Eels around the *Medens* somewhere." She spoke again in her very plain and unexcited tone. "Yes, pick your jaws up, it's true. The ship was programmed to look for egiya readings and we just found some. I'll have the *Medens* shifted around so that Observation Dome 3 can get a good look where the readings are strongest. The egiya readings are peaking so if you want to maybe see a one in million occurrence now's your time. I'll hold the bridge; the rest of you enjoy the sight, if it comes about."

Emmalyne put her hands around the back of her neck. "Ugor Eels. They've only showed themselves six times before; this is unreal."

"Seven times," corrected Hames. "They've appeared once while you were in stasis: a freighter leaving the Dover Space Docks spotted them. This time, out here if it happens, that'd make it eight." Hames had trepidations about the eels; it wouldn't be his first encounter. It was a rare experience, though, and one that could not be denied. "You want to take a look?" Emmalyne raised her eyebrows at Hames with a glance of questioning amazement.

"I'm glad you'll get such a great opportunity in your first week; that's a good sign if there ever was one." The exciting tidings from COO Melini Guerra were not matched by her tone and focus, which remained on the various outputs of data from the screens of the *Medens*. "Leena, let's reposition for a view from the middle of the ship around Observation Dome 3, thanks."

Leena gave a firm nod and returned to the pilot's seat. In front of her, a featureless block of grey matter shifted to allow openings into which Leena placed her hands. The grey matter then closed around her wrists, fully locking them in place. Then, from behind the pilot seat, a cord extended with a circular paddle that stretched to attach itself to the large golden spot on her neck.

Without request, Leena explained the purpose of her marking to Emmalyne. "Now you can see what the Sentir spot is for. I'll be able to guide the *Medens* more accurately, more efficiently, as it feels my directions." Leena's deep brown eyes narrowed as she began to bank the *Medens* upwards and to the right, following a tracking screen in front of her. The short movement was quickly over and she withdrew both hands from the grey matter and the circular paddle detached from the golden spot on her neck. The short and solid first pilot, now disconnected from the *Medens*, stood up ready to depart with Hames and Emmalyne. "Melini, are you sure you won't join us?"

"I'll watch from here. Too much to do and someone has to keep an eye on things."

Hames, Emmalyne and Leena departed at a fast pace, silent in the absence of any need to restate their purpose to reach the observation room as soon as possible. Their arrival to the large clear dome-shaped space was bookended by members of the crew ahead and behind them.

Leena stepped into the dome shape first, walking with her arms crossed. "I'm not sure I believe it, do you? The eels that is," she exclaimed to Emmalyne as they moved slowly towards the bottom section of the translucent structure.

Emmalyne, too, was hopeful and apparently open to a point of common ground with Leena, "Me neither. I want to believe; it would be amazing."

"Hmm. Let's see what happens," responded Leena as she gazed outwards to the expected area of activity.

Hames sensed Leena was still perturbed by the conversation of a few moments ago. He was relieved at the lack of an escalated altercation between the two young officers. Taking the opportunity to introduce Emmalyne to the others around her, he pointed out the crew members on the other side of the dome, all pressed hard against the glass wall. "The big guy is Francois, the small guy next to him is Dao and next to him is Temett. I'll get to the others later if that's okay." The three crew members heard their names spoken and briefly acknowledged Emmalyne before refocusing on the space outside.

"Great welcome, Chief," Emmalyne whispered to Hames.

"Don't worry about it. There's a mission and ops meeting tomorrow, you'll get to see everyone then."

Observation Dome 3 was like being in the top part of a glass ball. A flat white floor cut it off halfway and protected the crew from the conduits and cords that existed beneath the gangways. Removed as it was from the lifelines of the ship, the dome was detached and silent. The lack of background noise was finally broken by the small thin man in blue overalls with short black hair, who turned from the wall of the dome saying, "According to the COO the eels are about to appear."

"Thanks, Dao, we needed that confirmation," laughed Temett, his thick and wavy brown hair bouncing as he moved to face them all.

"You are welcome, Temett," Dao replied, nodding. "I must add that it has been nearly thirty years since space eels were first sighted and only seven confirmed sightings have been seen. Many of my people have sought them out – followed readings all across our solar system."

Emmalyne leant into Hames, turning her head away from Dao. "Who are his people?"

Hames moved around to match her position and spoke quietly in response. "Dao's from one of the outer solar systems. Remote mining colony of the Kuiper Belt."

Hames observed that Dao Mai showed no excitement or passion at this revelation that his culture valued Ugor Eels highly. He wondered if deep down there was some anxiety in Dao about this moment, Hames couldn't tell by his face, which was always still and unaffected.

"Thanks again, Dao." Temett produced a broad, disbelieving smile that seemed to cover his whole face. "We're across the history, rare sightings and so on. What do you think, Leena?"

The first pilot was distracted and moving quickly around to the free spaces in the glass dome. "Who knows? It's something different for us all to see, real or fake."

Hames continued the introduction of his new Mission colleague. "Everyone, this is Emmalyne. She's the new officer working for the Mission Team. She joined us a few days ago, just like you did, Temett."

"Hi Temett, I'm Emmalyne. I slept in the stasis pod next to you for nearly two years. You look only look a little tired."

Temett looked unsure how to respond to this statement. "Okay, thanks."

Emmalyne shook her head at herself. "No, I just meant that you're doing well. Sorry that came out wrong. I'm tired."

"Sure, sure, not a problem. I reckon I've come out of it pretty okay. Hey, what about them, Emmalyne, do you think the eels are real?"

Emmalyne considered the question. "I saw the video of them off the old PCD hub-station near Mars. They look beautiful."

Leena pulled a face of concern, her voice cautious. "They're also meant to be dangerous. I want to see them but I don't want them to cause any damage."

"The ship's energy field should protect us," suggested Temett. "It was only when they were provoked that second time that there was a discharge.

"Maybe." Leena peered deep into the black lit sky. "There's a lot of unknown experience with the eels. They're not really known to our Universe."

Emmalyne's previous problem with Leena was still in progress. "Doesn't mean we aren't linked to them somehow? That there's no connection?"

"You don't let up do you?" Leena, exasperated, moved herself away from the conversation towards another part of the room. Hames was impressed with Leena today; she was showing unusual restraint.

A deep male voice from the glass walls facing out to the Universe cut through the growing tension. "We are here in this room, which feels like we are standing in space, with the stars around us. If we see eels, magnificent, if not, c'est la vie, we are here together." All the expectant spectators in the dome had turned to listen to Francois, the tall, thickset coloured man. Hames had noted many times that while Francois was a lower level operations officer, essentially tasked with everything the others did not do, his immense size and age gave him an unignorable presence. Together with his lined face and grey hair this could have made him unapproachable, yet his almost closed eyes softened his look and he possessed an air of kindness.

Hames appreciated the pause provided by the dark burly Frenchman. It appeared all present were content to wait in relative peace. The group in the dome maintained a patient silence as they searched the space around them for some sign of energy, any sign of activity. After a few minutes, Temett broke the quiet. "It doesn't look to be happening. I've seen the vids – they could be fake. It's hard to think that after hundreds of years of space travel these things just came to life."

"My people believe the PCD cataclysm was a trigger in their evolutionary cycle, like a meteor hitting Earth: devastation and then life," Dao commented as he scanned the sky. "The first sighting of the Ugor Eels from the Russian Space Colony on Ganymede not long after the Cataclysm must have been a very challenging point in time for those people witnessing the event."

Hames thought to himself, *Only Dao could make first contact with another species sound like opening a jammed bioscanner.*

Francois joined the analysis of the eels' entry to their Universe. "The Russians were so shocked that the audio from the event just has them shouting 'energiya' over and over again. That data framework from the first encounter, the egiya readings, is what I believe the *Medens* scanners still use to track Ugor Eels."

"Hey, Francois, did you ever see anything like energy eels on your planet?" asked Temett.

"Not on Cernunnos, no, nothing like them," responded the large man as he turned his head upwards looking for the eels.

Emmalyne was surprised. "Oh, you're not from Earth?"

"Could you not tell from my size? I am somewhat larger then the required fit for the *Medens*!" Francois' belly laugh was so loud that it felt to Hames as if the whole observation dome was shaking.

"I was born on Cernunnos but I consider myself French, French-Cernunnos we referred to it back home."

Dao Mai cut in. "I find the pictures of Cernunnos beautiful in its order. It is a very strange planet to look at with its symmetrical layout of continents and water plots."

"Oui, exactement! Some people call it the Chess Board planet as the land mass is laid out in squares interspersed with lakes and seas. When I was home it felt like that – connected but separate. All the nations that settled there have their own polygon defined area."

"You've been there, haven't you, Chief?" Leena asked Hames.

A couple of times in my younger days, just to look at the launch stations for the particle drive ships. You know, to check coordinate calculations and see how they compared to other stations on Earth and Mermai. It was an easy place to operate and live given it's so distinctly laid out. The storms there, though, they weren't small affairs."

"Correct." Francois' big hand gave Chief Hames Naughton a big clap on the back. "Young Emmalyne, I can assure you that in your short life you have never seen a storm like an Cernunnosian storm. They make storms on Earth feel like the breath of your lover as it coolly flows over the back of your neck at night."

"Now that is a storm I could link into!" Temett clapped his hands together.

Francois steadied himself to explain the violent weather pattern. "To you from Earth, Cernunnos storms may appear violent, primal and entirely destructive. On Cernunnos, they are expected and regular; it's why it was such a good place to settle. The trees, plants and fruit all so strong and sturdy, and big!" Francois' large arms flew open wide, nearly scuffing the head of Dao. "The Cernunnos apple, it is not an apple but is the closet thing to an apple. It has an outside shell like a coconut but the inside is soft, juicy and tastes like the Crofton variety. Have you ever thought of having to crack an apple?"

"That's awesome." Emmalyne began to unconsciously rub her stomach. "What about the animals?"

"Nothing as ethereal as what we are all hoping to see here. The animals have skin like stone; there are not that many of them either, well, at least in type."

"Dinosaur dogs was what I saw mostly," Hames chipped in while looking about the dome for any sign of energy.

"Oui, they are more like dinosaurs, just not as big and very slow. Thank the Universe they are all herbivores as you do not want to get in their way."

After a pause, Leena echoed the thoughts of most of them in the dome: "We're not seeing much are we?"

Hames saw the collective disappointment and doubt although he did not share those feelings entirely. Many years ago he had not been left disappointed on Jerrim station near Earth's moon.

"Well, that's that," Leena moaned with a shrug of her shoulders.

The crew that had gathered in Observation Dome 3 started drifting towards the exit point, a move that suggested their departure was imminent. When, suddenly a green shimmer appeared in the space in front of them. Hames could feel their collective breath being held as in front of them green spaces no bigger then the size of one of their bunks began to form.

Emmalyne spoke quietly, almost to herself. "It is true."

Slowly, within the newly formed luminescent green space, long stripes began to turn an aqua colour producing a streaked effect across the agitating area. A short moment later crew of the *Medens* watching this event witnessed the elongated aqua aspects of the shining energy space reveal themselves as discrete and defined beings of energy swirling around each other.

Francois spoke for them all: "Yes, it has indeed happened. Here and now."

"Amazing," Leena gasped. "I'm still glad we're in here though." She shot a quick glance at Emmalyne.

"They absolutely seem alive." Temett looked in shock. "It's not just a bunch of random shapes. They stop and wait for each other then move again together."

The diminutive man from the Kuiper colony, Dao Mai, was in agreement. "It is unquestionable that this group of beings is alive and they have chosen to be so in front of us."

Emmalyne appeared happy and affected. "The way they pair up and follow each other, then separate and find each other again; it's a beautiful dance."

Hames decided it was an appropriate time to let her share a moment with her new crew members. "I'm going to finish up some of the mapping back in the control room. Let's call it a day."

"Sounds good to me, Chief." Emmalyne barely turned her eyes to him in order to stay focused on the events outside the ship. As he turned around and stepped into the darkness of a poorly lit exit point, Leena spotted his departure and joined him.

"Hey, Chief, where are you going?"

Hames wasn't expecting this. All the others were focussed on the clear walls, immersed in the amazing beings that swam next to the *Medens*. He spoke back, trying not to be heard. "I reckon I'll leave you all to it. Ugor eels, not really my thing."

Leena's dark eyes widened with confusion. "Chief, it's like seeing a comet that only comes around every thousand years, it's that rare."

"I've, ah, seen this type of 'comet' before."

The first pilot nodded in recognition. "Oh, you'll have to tell me about it sometime."

"Sure, not a problem. For now, though, I've got some work to do." He turned to leave but did not feel quite yet permitted to do so.

With arms crossed once again, Leena pointed with her thumb back to the room. "Emmalyne is quite new, I'm sure she'd want you to stay."

"She'll be fine, as long as you guys take it easy on her."

"Sure. Just don't forget, you're part of the crew too." She gave him a wink and turned back around to join the others.

Hames, alone at the exit point, watched the transfixed group. His own interest neither sparked nor captured, just memories triggered of experiences from years ago. He walked backwards out of Observation Dome 3 and took one last look at the Ugor Eels swimming in their fluorescent space suits, the intense and unbounded energy illuminating Emmalyne and the crew of the *Medens* in a glowing green shimmering light.

Chapter 3

Earth Before the Cataclysm

NISHA OPENED HER EYES AND closed them again. Lying on her bed, staring out across the rickety rooftops of a clouded south London, she dreamed of home. In her mind she sat on a brown sandy beach and the warm-red Mermain moon shone down on her deep olive skin. The picture was far clearer and more beautiful than the one stretching out from the silver-framed windows of her small studio. The beach fronts on Mermai dwarfed those on Earth; long tides and shallow waters gave an image of endless brown pastures of sand, only occasionally lapped at by a distant sea. She used to stay on the beach all day long, her heavy dark hair never dry after endless swimming. Finally, she would reluctantly take the long walk back up the stretching sands to her home or 'Ghara' as her parents use to call it.

Nisha arose and dropped from her bed with her customary thud onto the warm concrete floor. Unlike other females on Earth she was short and had what her grandmother called a guardsman's physique. This had never bothered her and the slight reduction in Earth's gravity, compared to Mermai, had lessened this appearance. She was more aware that her olive skin told all who saw her that she came from Mermai.

Nisha's mind began to awaken and she stretched her arms out towards the dispensary point. Coffee and yoghurt were requested with yawning between the stating of each menu item. She sat with a vague morning stare at the sky before her, taking in her breakfast absent-mindedly. Finally, Nisha turned her head away from the long grey clouds to examine a beige panelled robe. Her clothes for work at the Dover Docks were hanging ready for her, clean and ironed, a gift from her late entering partner the night before. Perhaps this was a peace offering for tardiness or simply because she was loved. With a cursory washing of her cup she drew on the dark navy pants and dark blue jacket with its horizontal strapping.

Ready for the day, she moved to the exit point of the apartment and ran her palm over a small square panel. Recognising her biosignature, the hard plastic door slid away and musty air brushed her face as she entered the corridor. A moment of trepidation registered in her spine as she braced herself for the corridor theme for this week. This was an unfortunate quirk of living at Battersea Towers in Battersea Park Road. The 1920s chic theme in front of her was tolerable. Gaudy chandeliers, gold-plated railings and rich red and white walls were interspersed with pictures of what she was told were "flapper girls". At least it was better than last month's theme, "Sailing Ships of the late 1700s". Apart from the battered wood walls, dank smell and biting cold, the wind that the programmers had set to greet people on their return from work was too much. Although it did achieve the intended objective of making her feel desperately happy when finally ensconced in her square featureless studio.

At the end of the corridor she entered a glass and steel elevator she was taken to the top of her building and onto an exposed square platform. As she was welcomed to the outside world, wind whipped the ends of her jacket straps back and forth against her sides. Nisha read a view panel that noted next week's theme for Battersea Towers, 'The Somme Trenches of 1917 France'. It sounded like a celebratory period, possibly a cooking theme. Another small surprise to occupy her time during her stay on Earth. "Only four weeks, just four more weeks," she said aloud to herself. Hames owed her big time for this although she would have followed him anywhere if it came to it.

A red rectangular transport craft with round edges and a clear dome coverage arrived and lowered itself to the platform. She boarded to become its fourth passenger and the vessel announced its take-off: "Departing Battersea Park Towers." Nisha grabbed a holding point in the seatless space. She felt a constant vibration as they were moved through the low sky above the buildings of south London. The first eleven months on Earth had gone slowly, in particular the first week, when she had wondered how she was going to thrive on this planet that seemed to be a never-ending city. Yes, there were pockets of green and certainly enough blue in the Earth's vast oceans. It was not like home, though, where buildings were a rarity. Just at that moment, almost as if the Universe sought to prove her wrong, the transport passed over a wide range of fields and trees. She knew the limited section of nature well as they passed over it every day on her journey to work at the Dover Shuttle Docks.

"Good morning, Nisha, you look lovely and healthy as ever." A pale, red-haired woman in similar navy work clothes moved from the back of the transport craft. A limp in her left leg restricted her middle-aged body as she attempted to adjust for the movement of the flying craft. Hers was a familiar face to Nisha, a friend from the outer colonies.

"Morning, Greta. You're still not having that leg looked at? I'm surprised that someone in the Lugurian counsellor core is not able to reset that bone properly."

Greta leant against the side wall a few metres back from Nisha, pushing her red hair back. "Oh, you know; it's hard to get a priority for expats from the colonies. Lugurians seem to struggle more than most."

Nisha moved to meet her limping fellow passenger halfway. "I forget sometimes how lucky I am here. Because of Hames we seem to get most of what we need. Many times at introductory meetings I have all these Agency reps thanking me very earnestly for my sacrifice as they note with gravity that Hames is a great mind and that humans are going to rely on him to take them past the seven colonies already established." Nisha pulled a comedic serious face as she relayed the explanation of her good fortune.

Greta laughed as she grabbed tightly to the holding point, swaying with the transport craft. "Earth is absolutely heaving with people and industry, and I know you don't want any more mass settlements on Mermai. So it's good to have your man on the job looking for more settlement sites."

"And I know Lugus has battled Earth for a long time to prevent any more settlements there." Talking to another person from the colonies was comforting for Nisha. There always seemed to be a feeling of togetherness in these conversations.

"In any case, the Eagleton Accord prevents mass migration," said Greta. "The seven colonies have their own constitutions, independence and migration policies. It sounds about right that the Agency is looking for more places to reach. I guess that's where we come in: you the shuttle docks officer and me the immigration official." They both nodded at each other. "Tell me, you could have chosen to do anything on Earth with a partner like Hames, why administration?" asked Greta.

It was true, she'd had choices. While Hames was notionally just a student at the Agency's Academy, he was on a scholarship and their lifestyle was fully paid for. She did not have to work and could have chosen a creative pursuit or a life of community service.

"Back home my artistic endeavours, while enjoyable, were not overly significant and my community work was, well, for my home, not Earth. So industry and administration feels the most habitual to me. On Mermai, industry was still the chosen practice for most people given how much development we still needed to do. It was a highly regarded choice." She looked ahead as Dover appeared in front of them. "The space dock at Dover; it always gives me a feeling of being linked to a route back home and in a small way makes me feel less homesick." Nisha raised her dark eyebrows happily at this simple form of logic.

Greta stretched her left leg against a base plate of the transport wall. "I hear you. It's why I work with resettling the incoming migrants from Lugus. I feel like I'm still part of home, working up at the main departure docks."

The red transport craft shunted hard, causing Nisha and Greta to hold tight as it flew down below a cliff top to skim above the water, spraying a net of mist about its path. The background of hard white rock served to highlight its pathway to the lower stations of the shuttle docks complex. Four tall pillars of chalk white steel and glass stretched upwards from the base of the cliffs and beyond their summit. Round and wide, it was only on approach that their enormity was evident. Waves licked the base opening of the second pillar as the small red craft eased into the landing bay, an act repeated by other transport crafts across the pillar bases. The craft's translucent covering and sides unfolded to allow Greta and Nisha to disembark.

Greta groaned as they walked towards the elevator. "You're lucky, your trip stops here. I have to head up to the shuttle dock and then the space dock."

Nisha wagged her finger at Greta. "Thank your stars you don't work on one of the main ships; it would be shuttle dock, space dock and then a trip to one of the colonies, and back!"

The limping transit worker winked back at Nisha. "What a day that would be!"

When the elevator reached her floor Nisha said goodbye to Greta and made her way into the administration area of the shuttle docks. Catching her eye, as it did most days, was the giant and brightly illuminated transport board that sat above the continuous corrals of workstations laid out in front of her. The screens showed the activity of all the shuttles coming and going from space, focusing on the transit platform sitting at the top of the shuttle docks complex. With a large circle shape in the middle and connecting arms stretching out to eight platforms, it was aptly named the spider. Nisha sought out her favourite piece of information on the board – the running total of travellers for the year. Already by June over four million people had moved in and out of the shuttle docks that year. The desire to travel by particle conversion drive to the outer colonies was not dissipating.

At the far end of the floor were the secluded work-pods for the Level 1 shuttle docks officers. Nisha entered an enclosed room with a plain grey desk and dark grey floor. The small office allowed limited visual stimulation to the officer within. However, behind her a full wall of transparent material provided a view that took in both the length of the Dover coast line and the horizon of the sea. Nisha considered this the best view in the house. Often during the day she found herself guiltily turning around to stare blankly out at the English Channel with its cold blue waves. As she placed her navy jacket on a hook, the computer read out her tasks for the day in its genteel voice:

> *"Finalise Performance Safety Reports to the SST Regulator*
> *Endorse re-draft version 11 of Safety and Operating Plan*
> *Endorse week 27 Scheduling Officer roster*
> *Review profiles for Induction – Officer batch 8248"*

Doing her best to ignore the expansive world of land and sea behind her, she gave some instructions. "Induction profiles first please." Welcoming new staff offered an excuse to take a shuttle up to the main PCD departure bays floating miles above her in high Earth orbit. While she scrolled through the profiles Nisha watched her director, Peder Jonslow, enter the work-pod to her left. The transparent wall that separated the two rooms provided them with the feeling of working together yet without any audible connection. Peder was a bull of a man, tall and filled out. His short blond hair sat up like a tuft of grass pushing out between sandy beach rocks and his voice was so naturally loud it had the effect of eliciting noticable anxiety in his team. Nisha always wondered if her diminutive but solid stature was a comforting counterpoint to all those who encountered Peder. Just as the director took his seat her screen flashed a personal communication message. Flicking the device on to see the message, she looked over to see Peder smiling hello. This smile turned to a half-grimace to show his disapproval of the promptness of personal contacts so soon into the working period. She knew it couldn't be Hames, he never sought her out during the day. It had to be Jasiah.

Her cousin Jasiah was more like a brother to her. When she had departed Mermai for Earth, Jasiah also took the opportunity to make a move away, to travel the Universe, "expansively" as he had told her. In reality he had not travelled much past Wandsworth Town and Clapham Junction, camping down between that short stretch of London. He had taken on various modes of employment while he produced paintings from his memory of home. Nisha had never understood this; how he had gone from being so determined to travel the Universe to becoming so fixated on one single spot. She suspected he may have grown uncomfortable with the idea of being parted from family and her presence in London was a reason for his stunted journey.

Typically, he would spend his day waiting for her to become available to keep him company. She did not mind this; Hames was busy most hours and the company was appreciated. For now Jasiah's call would have to wait. Peder Jonslow was sitting in the office next to her and she did not want to test his patience. Nisha denied the contact and started for the morning.

Her day progressed as normally as had most days in the eleven months she had worked at the shuttle docks. The passage of people across her office floor mirrored the arrivals and departures of shuttles to space, as shown on the large data board towering above them all. Excitement at the onset of her finish time of 3.30 pm was tempered by the fact that she was unlikely to see Hames anytime soon. He would be at the Academy until late and when he came home he would probably study some more, falling asleep with a book on his chest, a tiny blanket of numbers and calculations. Finally free from the day's work, she turned around to accept the sun shining in from the world outside. It spoke of warmth that could make even the grey and charcoal streets of London town inviting. The communicator buzzed again and finally she accepted the call from Jasiah.

"Hello, you."

"I knew eventually you would give in!" called Jasiah.

"Where do you want to meet up, my boy?"

"Let's head to a gallery down Carter Lane near St Paul's. I hear there's an exhibition of the tools and people from the Quadrant period of the late 2000s. I'm told it's gruesome! We can work up an appetite for rebellion!"

"Yeah, why not? Let's rebel together, I'll see you there at four," Nisha confirmed to her cousin.

Jasiah and Nisha exchanged a big hug when they first spotted each other. Like Nisha, Jasiah's hair was chocolate brown and nearly as long as hers. His skin was even greener then hers, which he regularly told her made him more Mermain. Jasiah's crisp image of white shorts and white t-shirt served only to highlight the picture on his chest of the Mermain moon, its red-brown hue shining out.

"Still promoting the homeland?" enquired Nisha with a smile and a laugh.

"You know it, cousin. People are blessed to see this image, are they not?"

Nisha chuckled. "Maybe. Let's go in."

Upon entry they were greeted by a sole curator standing at the doorway. "Welcome to the Quadrant Exhibition." The tall, thin elderly man waved them into a room clad with old timber panels with various pamphlets and posters on the walls. "Entry is free as this is Agency-sponsored and you are most welcome to look around, though I'm afraid I may be shutting the gallery before 4.30. The Agency has told me there is going to be a presentation around the corner on the steps of St Paul's." The curator turned as if finished but paused. "Oh, the presentation. It's going to be from the Head of the Agency himself, Legeles de Nord." He waved his hands around. "So you get the picture, lots of people, press, etcetera."

"Oh okay, we'll have a quick look and that should be enough," Jasiah called back to the curator while escorting Nisha into the main gallery room.

The rectangular room with the familiar white walls of a gallery held a main centrepiece of two large glass rectangular cabinets. Entombed within were various items such as electronic devices which looked like they sat on people's wrists and arms; black-hooded cloaks; and a montage of the symbol of the Quadrant movement, the four squares within a square, one black, one white, one grey and one red.

Many pictures around the walls showed wearers of the black cloaks with the embroidered four-square emblems. Some images showed the cloaked figures pulling what looked like office workers away by their arms, their legs dragging on the ground. Blood was a dominant feature in these pictures and relics.

Nisha and Jasiah looked at the images and objects for a while longer. Jasiah was content to view and move on to the next image within seconds; Nisha on the other hand took her time. Eventually she became fixated on a picture towards the end of the gallery of a young woman standing in front of a crowd of people who looked ragged and tired. They were all leaning against a stone wall looking entirely drained.

As she studied it the curator came up behind her. She turned to make use of his presence. "Can you tell us anything about the pictures?" enquired Nisha. "Who's that girl?"

"Yes, quickly though as I've just been told the presentation over the road is about to start." The curator stood next to the picture. "That was the final leader of a group that resisted the Quadrant priesthood and ultimately broke their powerbase. She's holding a head cloak from one of the four senior priests; she had just killed him. She and her comrades were part of the end of the Quadrant movement, and a good thing too. That lot were a nasty bunch intent on moulding us all to their image."

"Oh." The woman looked immense to Nisha. So small and petite amongst all those men, yet clearly their leader, their strength and their guide.

"Right, that's it, chaps. Time to close. You too should go, you really should. I'm sure you've seen the evidence of the work of the healers. It really is a change for us all; you know it's real when the Agency is backing it. You should go along and see what they have to say."

Jasiah and Nisha looked at each other with quizzical expressions on their faces and raised eyebrows, a double act for a single audience. "Why not?" they both said in unison, smiling as they left the curator. They made their way out of the gallery and started the short walk towards the front of St Paul's. Hot, humid air filled the afternoon, clinging to the old stone buildings that guided their steps.

Jasiah pulled at his t-shirt and shorts to let more air in to cool his arms and elbows as they became part of a growing population on the way to St Paul's. "It's a bit dense around here."

"I know what you mean; there's a real crowd forming."

They reached the back of the group and found themselves standing on the pavement in a closed-off street that was full of people. Nisha could see all avenues in and out of St Paul's had been cut off, ensuring a place for a large group of people. As soon as they came to a stop behind the crowd more people filtered in behind them. In a few moments they turned their heads around to see that they were in the middle of a throng with no discernible exit.

"What just happened?" asked Jasiah incredulously as he wiped sweat back through his loose brown hair, his armpits showing a wide patch of moisture.

Nisha laughed in confusion. "Ah, I don't know. I didn't know the Agency was going to be doing anything. I don't follow Earth politics."

Jasiah turned his head back and forth. "I don't follow Mermain politics let alone Earth's." Nisha and Jasiah were stuck here and it seemed an effort would be required in order to leave. Standing next to them was a slim grey-haired woman wearing a blue suit who looked intently over to the steps in front of St Paul's. Jasiah caught her attention with a half-hello wave of his hand. "Excuse me, I don't mean to sound vague but do you know what is going on here?"

"Of course, the Agency is realising the findings of the Cetana Initiative," the blue-suited woman responded in a matter-of-fact manner. Nisha's vague shrug in response prompted the stranger to comment dismissively. "Oh, I can see you're Mermain. You might not know or understand Earth's perspective."

Nisha's brow furrowed as a hint of frustration welled within her. "That might be the case. You wouldn't know though."

The woman returned her gaze to the steps, seemingly expecting the arrival of the promised speaker. "The Eagleton Accord may have left the colonies with self-determination but it left Earth feeling disconnected. We are disconnected."

Like most people from the colonies, Nisha understood the arguments and issues for breaking free of Earth. She had celebrated wildly when the Eagleton Accord was signed. "So, because you don't own us anymore you feel lost?"

The sarcasm was registered by the woman in the blue suit but she didn't respond. Perhaps it was due to the heat, perhaps it was the woman's strength of belief. Regardless, the lack of sweat on the person next to her frustrated Nisha greatly, nearly as much as the tired arguments of exo-planet ownership.

The grey-haired woman offered one last comment without even looking at Nisha. "My young Mermain friend, the Agency has a plan for us to become connected again. To reach the goal we all want. To understand the purpose of all of us, you know, humans! How we can stay together. Finally, the Agency has something to say about it. To set us on the path, to allow us all to get there!"

"Get where?" Nisha asked leaning over. She was about to ask another question when a man's deep voice began to resonate through the speakers set up at the top of the steps.

"Thank you all for coming here today. Today is a day we mark history and stop marking time. Today is the day we move on from the squabbles and prejudices of the past. We seek only to know what we already know, that while we are bound here on Earth so too are we bound to the sun, the stars, and the whole Universe around us." As the speaker finished this statement cheers rang out around the crowd, suggesting to Nisha that these people knew as much about the topic as the speaker did. People called out, "Tell it, Legeles", or "You are with me" and "We are with you".

Jasiah turned to whisper to Nisha, "I'm not sure about this, Nish, it's the healing stuff, yeah? Something weird is going on here. It doesn't feel good to me, anti-colonist I reckon."

"Let's stay and listen," she whispered back, curious as to the purpose of the event.

The long white-haired speaker, Legeles de Nord, spoke again to the crowd. "From this point, the Agency and I are sponsoring a program of healer training and a program for those who want to take the step towards inner growth. For those who want to take on love, not conflict, as their motivation. We will provide places for meditation and meditation training. We will ensure no-one is left behind."

More cheers rang out with various calls of approval. Nisha was slightly impressed even if uncomfortable. She whispered to her cousin, "You have to admit, Jas, it sounds alright. I like meditating. It seems more like the community intent I know from Mermai."

Legeles de Nord continued. "We have touched the Universe through our travels to worlds far away yet we have lost ourselves and our harmony. We need to all be as one, in order to move forward as one. Over the years Earth and its people have been fractured through its stretch across the stars. Now our people need to realise we are all together across matter, across time, across minds. Earth must be united and live as one!"

As more cheers and jubilation rang out, Jasiah's hand grabbed Nisha's elbow. "Um, but we're not united with Earth, we're Mermain. We're different." Nisha shot a look at Jasiah to indicate that this wasn't the best time to raise that idea, even though their deep olive-green skin clearly gave them away as non-local inhabitants.

Legeles de Nord brought his hands down to calm the crowd. "To this end, I want to also announce the intention of this Agency to no longer support the constructs of the past. The last remnants of religion have been propped up by the State for too long. Here on the steps of St Paul's today, I announce the end of that financial assistance."

For the first time, Nisha thought she heard one or two negative noises amongst the more muted cheers, although these did not persist.

"We will not actively stamp out these beliefs, for that is an individual's right. But if these practices cannot be self-sustaining in our society then they have no place!"

The emphatic nature of Legeles de Nord's last statement brought back the people's positive calls and Nisha could hear only the euphoria around her. The woman in the blue suit was clapping wildly and crying. "Isn't it fantastic!"

The strong comforting tone of the main speaker calmed the crowd. "As you all know, we have concluded the ten-year long affirmation of oneness. The global Cetana Initiative, through its work across all lands and with all peoples, has proven beyond reasonable doubt that our connection to each other is there, just simply veiled and held back. It is time to liberate ourselves. To free our consciousness from its localised boundaries and to reach out to each other with warmth and compassion. To reach oneness with a liberty of consciousness!"

The last statement brought with it a push at Nisha's left shoulder from arms and hands behind her reaching upwards in jubilation. Once more, Nisha and Jasiah were caught offguard by a thickening and deepening of the crowd around them.

Legeles de Nord moved on. "To reach that ultimate goal of simply knowing you are free and able to do so much, to be so much, we have a program of meditation, healing of the mind and, ultimately, steps to move beyond the known gateways of consciousness. This is our program and this is your Universal Direction!" At this final shouting statement banners unfurled to reveal Legeles de Nord's final words of "Universal Direction". Underneath these words was a symbol of eight circles, all linked together to form a ring of eight, all a golden yellow colour that seemed to shine brighter then the sun.

Nisha spotted banners dropping all around them to repeat the name and image. It seemed the more the banners dropped the more the crowd's fever rose.

Calming the people with his open palms, the white-haired leader pushed on. "From the middle of this year we will be moving to establish a Universal Direction program for all. A program that can be accessed any time, any place, and at..."

Nisha could not hear the rest of the sentence. Her head was suddenly light and her ears were ringing. Something had happened on the far side of the crowd, causing people to turn away from the speaker. Then, Nisha watched people beside her fall down as a tall young man in plain brown work clothes pushed his way through the crowd. She could see that this man was angry; his long arms flayed bodies away from his path. Glowering, he gestured aggressively to Legeles de Nord while he pushed more and more people aside. The man moved clumsily; something protruded from his dirt brown work shirt. When he reached the steps of St Paul's, one of his feet caught the other, causing him to trip, fall and explode in a moment of dust and devastation. Nisha was blown backwards by the explosion. Her last thought of the angry protester as her consciousness ebbed away was, *That was an unnecessarily hard fall.*

Chapter 4

Mission and Operations

CLOISTERED IN THE REPAIR CONTROL room in the far right wing of the *Medens*, Hames spent the morning familarising Emmalyne with the progress of repairing the Mermai Tentacle, without much success. To his frustration he could not get her to focus. Whether it was the constellations shining down from the clear round ceiling or Emmalyne's incredulity at the shape of the *Medens*, their educational headway was as inhibited as their expedition to repair the broken space ahead of them.

"Chief, you have to admit it. The *Medens* looks like a big flat ray-fish with its wide body and gaping mouth," Emmalyne asserted.

For reasons unknown to him, Hames did not want to give in to this argument. It was true; the *Medens* did look like a giant grey-green manta ray, especially with the noticeable lengths protruding from the mouth of the particle collector. "I don't have to admit anything. That wide gaping mouth is the particle debris collector, it's meant to be that way."

In this dispute, all concentration on the task in front of her had departed Emmalyne. "Even the back-end of the ship, where the converted debris comes out, well, that could look like a fat fish tail." She nodded her head in affirmation. "The designers did it on purpose. Yep, we're agreed." Emmalyne turned away from Hames, signalling her view that this argument was settled.

Hames was saved from further disagreements by the familiar three-tone crescendo signalling that the regular meeting of the Mission and Operations groups was soon to commence.

"What's that for, Chief?" Emmalyne frowned.

Hames moved from the bench covered with screens and translucent galaxies maps, heading for the exit point. He threw over his shoulder, "It's nothing to worry about it, just the Mission and Ops meeting. These happen every eight days or so." The exit point behind him reached its full opening and a small bow wave of sound and warmth washed past him, heading towards his hesitant colleague. "We've about fifteen minutes to get down to the mess hall." He beckoned her to join him. "We've got to get moving, okay?"

"That's on the other wing of the ship. I would have thought we're better served here."

The chief mission officer shook out his strong arms and shoulders, flexing them behind his neck in a satisfying stretch. "Hmm, I've never considered whether they're mandatory. Never hurts to say hello to people though. C'mon, you can get that formal introduction I promised you."

For the first time that morning, Emmalyne faced up to her screens and charts, a newfound dedication apparent in her body. "Since the *Medens* stopped to pick me and the other new officers up, we've not been travelling at full speed. The Ops meeting has to be about moving back into light alignment mode, the scale-up of the repair engine output. The meeting will be focused on these tasks, not new recruits."

Hames could see her discomfort about the looming meeting and the potential degree of focus placed on her. This was slightly confusing after her complaint the other day in the observation dome that she didn't get much of an introduction to the others. "It's a good chance to see your team. You have to take these meetings seriously. The Mission Team is responsible for delivering on the primary purpose that the *Medens* was built for – fixing the PCD fractures. Like you said the other day, repairing space is a powerful purpose."

Hames said the words "powerful purpose" although he didn't feel them as a truth within himself. Yes, the *Medens* was performing an important job; the task to repair space was critical to helping humans travel among the stars again. It just wasn't the full reason he was here. His was much more personal.

Begrudgingly, Emmalyne followed Hames on the path that would take them towards the meeting. The grey-panelled corridors and gangways zigged and zagged much like a computer relay. The path appeared so oblique that it may have told a person new to the ship that they were in fact walking without direction, lost in a constant maze of homogenous halls. Although he had walked the paths of the *Medens* for three years he still found them difficult to remember. His focus was always elsewhere, not on the immediate step in front of him. When they reached a choice in direction, Emmalyne brushed past him, her body more certain than his as to their destination, her silent striding ahead of him a marked change in approach from just a moment ago. He wondered whether Emmalyne would continue to have challenges ahead in adjusting to reality after dreaming for so long.

At the middle point of the ship, near the repair engine itself, they saw ahead of them Dao Mai, the clear-speaking man from the observation dome. Hames motioned forward with his hand. "There's Dao working on the panel next to the exit point. He's the mission technical leader, I'll introduce you."

The man ahead of them caught sight of his two colleagues and turned with a slow and purposeful manner. "Chief Hames, I have just finished the upgrade on the communications conduit from the control room to the repair engine room. At 9.32am I realigned the insulators to ensure connectivity to the back-end systems and patched through..."

Mid-sentence Hames held his hand up. "Thank you, Dao, that's fine."

"You should now have the scanning capacity for reviewing previous journeys along this PCD route by other PCD ships."

Hames's uncomfortable smile reflected his preference to not share this information with Emmalyne. The upgrade performed by Dao was, in practice, not fully essential to the operations of the *Medens*. More so, it was an enhancement that would provide him with information about which ships in the past travelled this route and, possibly, when they travelled this route in the future. All to help him in his own personal mission. Unfortunately, the expressive and verbose nature of Dao Mai had brought him undone. As a point of distraction and purpose, Hames considered it best to focus Emmalyne and Dao on their introduction. Prior to his attempt to alter the subject Dao continued his descriptive outline.

"Yesterday at 10.47 am I finished inducting the two new members of the Operations Team. Their names were Temett Bakker and Robero Paldini. I do confidently believe that they will be adequate in their roles and understand their limited access to the critical mission areas."

"That's great work. Dao, this is Emmalyne Biggs, she's one of the new crew members." Dao Mai said nothing and simply looked back at

Emmalyne. "Are you on your way to the meeting?" Hames asked in a formal tone.

"I am soon to be walking through the modules towards the Mission and Operations meeting, yes. In two minutes, I will finish collating my tools and will fold the mesh envelope, clicking the seal before I stand up and move towards the meeting."

Hames realised that Dao was uneasy about meeting Emmalyne for the first time, as would be many of his people, and he was falling back into old and bad verbal habits as a coping mechanism. "Okay, that sounds fine," confirmed Hames. "Emmalyne and I'll walk ahead. You catch up when you're ready."

Hames and Emmalyne moved on from Dao Mai and not ten metres away the discussion turned to her new colleague. "He sounds like an astronaut." Emmalyne frowned at Hames.

It was a completely true statement. Dao Mai looked and talked every bit like an astronaut, which unfortunately in Hames' lifetime had grown to be a derogatory reference for small groups of people who lived permanently out in deep space. Dao had grown up in an outer system mining colony. This was an atypical life for astronauts yet Hames was reluctant to stereotype these people, or anyone.

"Listen, Dao's an extremely intelligent guy and a very important member of my team as technical lead. You need to keep that in your mind when referring to him."

"I know, but they speak so formally. I feel this instant wave of boredom come across me. I'd rather be in stasis than be around them for any length of time."

Hames and Emmalyne continued to walk side by side although he didn't look her in the eye when responding to her last comment. "Yep, space colonists. I know they can be tough work. For them, though, words and life operate differently. They have this kind of weird patience that comes from growing up in the dark of the Universe. They feel time in another way compared to us. Not many people on Earth understand that." Hames finished this sentence by gesturing with both hands downwards to try to firm this view to Emmalyne as final.

"I don't know how kids can grow up like that. Living all their life with just a couple of people in a colony so far from Earth, farming nebula gases or mining rocks in exchange for food and water. It seems so feudal to me." Emmalyne scrunched up her face.

"I think that's why people chose that life. It's quite simple and predictable. I guess the only problem is that over time that simplicity combined with the dangers of space created a very cautious culture. The need for clarity created a very, well, um..." Hames paused to acknowledge her previous comment. "A plain type of language and person." Hames had a great deal of regard for Dao. Emmalyne did need to show her colleagues more respect. Her lack of tolerance was surprising, possibly a product of coming out of stasis.

"Sorry for being blunt. I'd never thought about their point of view much before. We're still connected to them after all," tendered Emmalyne under her breath.

Soon Hames and Emmalyne found themselves in the corridors of the front module and at the mess hall where the meeting was to be held. Warm and bright, the square-shaped room had colours that reminded Hames of home with light-green panels, a sand-coloured floor and a blue ceiling. Hames appreciated the design intent, to pacify and remind him of Earth; it did make him feel calm and content.

Emmalyne paused at the entrance point. "When I came in on the transit ship I noticed Temett in stasis and the other guy, Robero? They're the other new crew members that Dao mentioned. I think they were waking up like me? It was such a daze, pretty hard to remember it all."

Hames motioned her in. "They should be here. Probably feeling a bit out of cycle just like you."

Emmalyne's face searched around the room. "Okay. It's good to know there's others starting fresh like me."

The mess hall rang with life as the crew congregated around a large white rectangular table. Some groups of two or three were analysing forms of pastry from the food dispenser. Some were still talking excitedly about the experience the day before with the Ugor Eels. Others were already seated and ready with drinks and food to tide them over during the meeting. Hames' typical focus on the *Medens* was predominantly internal yet even in the mess hall, the smells of rich dark coffee and freshly baked bread wilted his contrarian resolve. He could not help but feel relaxed as he led Emmalyne towards a seat in the middle of the table.

In front of them, a young man with tired and blank eyes, turned to absently drag his arm across the table and knocked over a cup of water. For a brief moment, the unbridled water started to spread; then it stopped and held its position before retreating and forming a rough circle position in front of the fallen cup. Unsurprised, the young man looked around the room and opened a nearby wall panel to grab a white piece of cloth. After he dropped it on the water, the liquid was instantly mopped up by the material, which seemed to expand and contract on its own.

Emmalyne had watched the event unfold. "I think, um, he was one of the two others on the ship with me."

"Yes, that's right, Robero I believe," Hames confirmed. "And you've met Temett, he's over there." Hames pointed to the tall man sitting at the other end of the table.

"Oh, yes, I remember him from seeing the Ugor Eels, and the stasis pods. I couldn't forget him and his freckles; they look so out of place."

"Probably not a great observation to make to him in person. Just my suggestion," Hames cautioned her although she was right, Temett was distinctive, and confident. He had a permanent grin which could, in the short time Hames had known him, stretch to an all-too confident smirk.

Melini Guerra entered the mess hall, silencing the environment. The room seemed to follow her as she moved to the middle of the table and sat down. As she sat the seat beneath moved forward and, almost unnoticeably, warped around her to make a perfect ergonomic fit to her solid Mermain stature. Her head bowed slightly and her long, thick hair, a mixture of grey and brown, dropped around her head. She tapped two buttons on the table in front of her, and the space morphed ever so slightly to make an even crack which then moved further apart to form a small screen.

The COO looked up with a serious face. Her brown eyes were big and strong and her eyebrows seemed permanently pushed down, making her visage all the more serious. "Alright, let's get this meeting underway," Melini Guerra announced. "We've been coasting for the last few days since we took on board the new recruits. Officially to the ship, I'd like to welcome to the Operations Team, Junior Engineer Robero Paldini and Second Pilot Temett Bakker. To the Mission Team, please welcome Emmalyne Briggs, Junior Mission Officer. They are all, as you would personally remember, still disorientated from their time in stasis so keep an eye on them. Lower your expectations as it will take a while for them to come up to speed."

Hames scanned down the agenda for their discussion today; nothing too unusual. He did expect the odd moment of conflict, which was normal for a ship so far deep into space, although the COO was adept at not letting this type of behaviour go too far.

A man with grey curled hair and a large belly moved from a dispenser wall and handed the COO a steaming mug, which most of the room knew was extra strong peppermint tea. The officers to her left and right began to recoil as she sipped on the fragrant liquid. The ability for Melini Guerra to consume this drink without flinching had always amazed Hames.

"Thanks, Siggy," exhaled Melini Guerra.

In his faint Scandinavian accent the ship's manager responded brightly, "You're welcome!"

The COO clapped her hands together. "Right. Now we've had the excitement of the eels pass it's time get on with operations. 'Item 1 – Move to Light Alignment Mode'. We've been in cruising mode for three days now. The periodic shutdown of the light alignment drive is nearly done. Tomorrow we move back into LAD mode and get moving. Yolanda will coordinate with Chen Chung from the Mission Team so the repair engine and light alignment drive transition smoothly together when we're ready." The COO acknowledged a tall woman with short-cropped white hair sitting to her left.

Emmalyne whispered to Hames, "Who's that?"

Hames answered through the side of his mouth. "Chief Engineer Yolanda Hollings. She looks after the light alignment drive."

The COO continued. "Before we move on, does anyone have any issues with the transition?" Silence greeted her question, which Hames believed was a statement of the crew's ability to comply with her orders. Most of the crew.

Leena raised her arm and began to speak before it was acknowledged. "I'd like to get to the update, efficiencies and progress please."

Yolanda spoke over the table to the ship's first pilot with amusement. "Always with the progress report. You're as regular as Melini's choice of tea."

The COO accepted the request as if it were the first time it had ever been presented. "Sure, Leena, let's do the update now." Melini Guerra tapped at the screen in front of her and brought up the route map to Mermai to hover in bright blue lines above them all. Like the line of a fish in the deep black sea, it showed the completed repair trail from Earth and the remainder towards their destination. "We've repaired about 12.5% of the Mermai Tentacle in total and we're meeting nearly all of our key mission indicators."

Leena's small shoulders leant inwards to ask her next question. "And the efficiencies?"

"As you know, they are some of the indicators we're not reaching," responded the COO.

Yolanda confirmed this. "It's been slow progress with increasing our speed. We're going as fast as we can, for now."

Leena played with the golden spot at the right of her neck and then closed her hands together on the table in front of her. "It's still so far away; not much different from the last few meetings."

Yolanda leaned sideways towards her leader as she answered the statement. "We've been out of light alignment mode for over a week now, it's understandable."

"If we keep going at this rate, it will be twenty years before we reach Mermai."

Melini Guerra brushed her hair past her olive-skinned forehead. "Leena, that's the overall goal, not exactly our goal though. We're here to start the work on repair on the tentacle. There's no expectation we'll complete the job."

"What's our turn-back point?" Leena's hand came down with noticeable noise on the table.

"I've gone over this before; we don't have a set return point."

"At some stage you have to firm this up. I know there's harmony and connection in us all. No matter where we are. Here, now, the past. I just want to make the most of this experience, and I don't mean life on the *Medens*."

"Okay, enough, Leena. We'll talk about this in the next few weeks." The COO's voice revealed a touch of frustration in this comment; it was fairly obvious that Leena's issue had been registered, time and time again.

The COO looked down at the screen and did a quick double-tap with her pointer fingers on the screen in front of her.

Hames drifted in and out of the next few items: discussions over the use of ship supplies, changes to recreation schedules and an update from the head office for the Organisation back on Earth. He spent these moments thinking about other times, other times from long ago. Thoughts of her, alone on Mermai, or maybe not alone. Maybe she had moved on? It was a question he had never been able to answer properly. He looked down at his hands, this time focusing on the broken compass, his reminder as well as his communicator. A thought occurred to him then and there, *She has never seen this, this broken image.* He had carried it on him for so long, and it only struck him just then that the weathered picture had never been seen by its subject. Was this a waste of energy? Most people moved on or healed after going through the Cataclysm, but not him. Even in the relative twilight of his life, he was still working, working his way back to Mermai.

Hames brought his mind back to the meeting, focusing on the words of Melini Guerra in order to realign with the present.

"Okay, what you are all waiting to hear about. We're only five weeks out from reaching New Galapagos. Let's not get too excited about this; as much as you have this feeling like it's a holiday for you, I can tell you it's not." The COO tapped some buttons in front of her, bringing up in the middle of the large white table an image of four moons surrounding a lonely, small, rock-like planetoid in space. "This system is the oldest post-solar system colony from Earth and the only one we can reach without particle conversion tech. Remember, it hasn't been seen by anyone from Earth for decades."

Ship's Manager Siggy Conson continued the explanation of their pending event. "Trade with New Galapagos dropped off a long time ago and communications in the past few years have been sporadic, at best. We believe the Administrators are still keeping good control of the citadel but we don't know much more than that."

Melini Guerra touched on a more procedural aspect of their planned visit. "There could be a range of issues and feelings that we just don't know about. So it will be a cautious visit, at the least for the first few days."

"Is it compulsory to attend?" asked Temett with his half-smirk smile.

"I would have thought all of you would be begging for shore leave," the COO responded. "I know you all signed up for this work with no intent at being public relations officers for the Organisation. Our visit is important for the people of this system. It's one thing to be a small group; to know you are a community cut off from everyone else is another thing. These people deserve our time and I know all of you will welcome the change of scenery. If you do have any feelings of anxiety about the visit, Healer Olisa has made herself available to help with anything that's troubling you." The COO motioned towards a scruffy and unkempt woman sitting a few seats away from Emmalyne. The woman appeared older than all of them put together. She sat arms folded; the Universal Direction symbol, buried into a leather band on her right hand, spun as they all looked at her.

Hames winced at the idea of a heart-felt conversation with Olisa, about himself or Universal Direction. He attempted to think of more pleasant things to counter his facial contortion, *chocolate and ginger, melted in a hot drink pouring down my throat.* This thought was over-ridden by the voice of Siggy Conson. "I'll circulate some information closer to the time. Where we'll be staying, what we'll be doing and so on." This information was met with an array of nods and smiles. Finally, Hames heard the familiar sound of a low slow bell followed by a slightly higher measured slow chime. A welcome sound for some as the Universal Direction session for the day was soon to be held in the quadrangle.

"Okay, it's time. If there's nothing else then that's it." Before Leena could raise her hand the COO walked out of the mess hall. The Mission and Operations meeting was over and the greys and blues made their way out of the Earth-coloured room.

Hames Naughton pushed away from his chair and felt his beard, watching the others leave the room. "That's it for the day, Emmalyne, and there is still 'day' out here you know."

She sighed. "I can never tell, my body clock never seems to adjust for space."

Olisa smiled at them as she left the room, he strained a grin in return. Blatant politeness was the best avoidance technique he could employ. Olisa had been seeking a meeting with him for some time, which he had avoided ever since she stepped onto the *Medens* years ago. Given their journey of repair could stretch on for years he was unsure this ability to avoid the healer would last forever.

"There's some nice folks here, some strange ones as well. Especially the astronaut," Emmalyne commented in far too close proximity to Dao Mai for Hames to respond. "Anyway, I'm off to the session. See you, Chief." Before he could respond she was gone.

Despite her struggles with adjusting to her new surroundings, Hames had to admit he liked Emmalyne. The uncomfortable situations she created made her seem different and more alive than the others. She reminded him of a different time in his life.

The remaining crew abandoned the mess hall and he felt detached from the group. The quietness was interrupted once again by the ringing of low bells and chimes that he felt vibrate in his chest and out across the walls and floors. The vibration was a call to people on the ship, a call to Universal Direction he was not going to answer, not now and most likely never.

Chapter 5

Light Alignment Drive

THE MORNING HOURS DISSOLVED as Hames worked in the repair control room alone. Emmalyne was absent again without explanation and his dangling solitude was only heightened by the electronic sounds of check and confirmations. The *Medens* was readying itself to enter into light alignment mode and increase the progress of repairing the broken Universe ahead of them. Hames knew she would not want to miss this and eventually, with half the time left in their shift, Emmalyne finally joined him.

"Chief, I'm sorry, I got, um, distracted. I was talking to others in the home-pods about the eels from the other day. Wasn't it amazing! You could almost feel them as much as see them. It was like, like energy that was alive."

The chief mission officer's brow furrowed all the way down to his beard, deepening his voice. "They're attractive to look at, sure. But they can also cause trouble. Do you know what happened outside the Jerrim Transit Station between the moon and Earth?"

"After Leena mentioned they could be dangerous I read up about that incident. They tried to grab one with a Faraday cage but it emitted a pulse that shut down the whole station." Emmalyne's excitement reduced somewhat.

At this point Hames didn't want to share his own personal history of seeing the eels and his very presence at Jerrim Station. Right now, Emmalyne was intensely distracted by the egiya beings and the ship was soon to undertake one of its major operational manoeuvres.

Hames started to pack up and close down the repair control room system. "It was good fortune that only a few people died on Jerrim. The survivors were really lucky to make it back down to Earth. I don't want to spoil the eels experience..."

"...But you will go ahead and give it your best shot," Emmalyne finished for him.

"There's a lot we don't know about them so perhaps cautious curiosity is the best response if we see the eels again, not awe." Hames powered down more systems.

"Yes, yes." Emmalyne stretched her arms and fake yawned while she watched Hames finish his chores.

"Let's head down to the repair engine room. We don't want to miss the shift into light alignment mode." Hames closed down the conversation of Ugor Eels. It was time to concentrate on fixing the space ahead of them and making a path to Mermai.

The sound of their steps on the gangways below was broken by the curiosity of his new colleague. "What's the story with Leena?"

"What do you mean?"

"She seems to speak up more then the others."

"I think she's pretty focused on seeing things are right."

"One of my old instructors from the Organisation was of the Savai branch. He was always trying to get us to join him on these expeditions like rock climbing or rafting."

Hames considered this. "I never paid much attention to the different parts of Universal Direction. I think Leena being part of that branch might be why she's outspoken. It's hard to live life to the fullest if you're stuck out in space."

"There seems to be something more to it than that. No-one else spoke up against the COO or challenged anything. The rest of the crew was really respectful, not Leena though."

They turned a corner and the neutral light bar that ran at head-high height along the *Medens* corridors switched to a pulsing orange brightness, signalling the impending move to light alignment. Hames acknowledged Emmalyne's astute observations of the COO and Leena. "You saw that quicker than most. There's a connection between the two of them. Leena is something like a third or fourth cousin of the COO." He paused to consider whether it was appropriate to share any further detail. "Their connection is tenuous; that slither of relationship, though, is just enough for Leena to feel she can push the boundaries with her relative.

Leena's also a special case. She's the first pilot and was the only pilot until Temett arrived. It's been a challenge for the COO waiting for the pilot back-up; it's meant she's had to persist with Leena. Temett's arrival will help fix that."

Emmalyne reached the entry point for the repair engine room and in detection of her presence the space dispersed to its sides to reveal two officers at a desk with two screens in front of them. Leaning back in their chairs the men did not appear to be overly busy. The relaxed nature of these crew members was already known to Hames although not to Emmalyne, who whispered as they entered. "Look at them! You'd think they were asked to switch the light on, not move the repair engine into parallel with the light alignment drive."

"It's all an act," Hames spoke back in a quiet tone. "They're stressed right now, trust me."

"Sure, real worried," replied Emmalyne as she looked over to the empty desk space on the workstations.

Hames couldn't argue with her. He knew the easy demeanour of the two officers grinning towards him was at odds with the aura of the repair engine room. It was a place that made new entrants to it feel both completely lost in its operations and in awe, almost fearful of the critical nature of its function.

"This is Second Mission Officer Xavi and First Mission Officer Chen Chung; they weren't in the Mission and Ops meeting the other day. Have you caught up with them yet?"

"No, no I haven't, Chief." Emmalyne angled her body towards the two officers but did not make direct eye contact. "Hello, it's nice to meet you both."

"Hey there, Emmalyne, welcome aboard the *Medens!*" First Mission Officer Chen Chung grinned as he and his colleague Xavi shook her hand.

Emmalyne Biggs looked lost in analytical thought. "They're a bit like peas in a pod: same height, same short dark hair, same light-brown skin. Xavi, he's from the Southern Americas and Chen's from East-Asia," she observed very casually.

Hames felt a small crunch of embarrassment again. The statement of their ethnic ancestry was unnecessary. This was bearable, however, as he knew it wouldn't bother either of the two officers in front of him.

Chung laughed, almost brushing aside the observation. "We all go and come from the same place don't we?"

"You don't seem to have a care in your minds, do you?" Emmalyne stood at the rail of an elevated gantry that overlooked a large rectangular room with no windows, except for an observation dome set in the middle of the ceiling. Hames joined her at the handrail and they looked down onto a sparse room that held a large grey rectangular shape that was higher than most men and at least eight metres long and two metres wide.

Xavi stood up and joined them. "What is there to worry about?" he asked as he stopped next to Emmalyne. "We've done this lots of times."

"Look at that thing." Emmalyne was on the cusp of being lost for words as she pointed at the large grey block in the middle of the room below them. "The interface for the repair engine, right? I've never seen a real one. I've worked on them in labs, that was it."

"Yes." Xavi's bright brown eyes shone as he waited for more.

"And look at your desks, nothing on them. Nothing on the repair engine interface, just blank pieces of plastic and a few screens! These interface systems are the most complex systems we can work on!"

Chen Chung moved next to the others looking down at the large rectangular shape and shrugged. "Well, that's all you need to know about repair engine interfaces, they're complex! Wasn't it my job to explain what we do?" Chen Chung and Xavi looked at each other and laughed.

Emmalyne began to gesticulate at the shape below them. "Shouldn't you be down there somewhere? Pushing panels or something, looking worried?"

"Not really." Chen Chung grinned. "It would be impossible for us to manage the rate change in how the repair engine takes in the broken particles and corrects them. We'd have to escalate it just at the right moment the ship moved into light alignment and light speed. It's impossible for a human to do, so the computer does it for us."

"What about the layering system? Do you control this from the interface block?"

"You're right on that one," commended Xavi. "The layering is more about making sure the repaired particles are distributed evenly. Come down and I'll show you some more functions."

The four members of the Mission Team made their way down to the repair engine interface block, their footsteps tip-tapping on the plastic stairway. As they approached the large grey box a powerful hum registered in Hames' chest. He felt like he was at the heart of *Medens*, power pumping in and around the ship from this one control point.

Xavi turned to face Hames and Emmalyne with excitement in his voice. "Okay, you know it's not good for the repair engine to stop fixing broken space. The damage is all along the particle tentacle to Mermai and if we don't clean it all up any future ship using a form of clean PCD travel would just disintegrate. So it's important that we have a process for transitioning the activity of the repair engine as we move from cruising mode to light alignment mode."

"Yes, I understand that; the amount of space to repair gets exponentially greater once at light speed." Emmalyne rubbed two fingers against her right temple, looking strained.

Chung allayed this concern. "It's not a problem. The conversion process is adequately expansive in that it's almost instantaneous."

"Can the repair engine cope with that change in speed?" Emmalyne understood but her voice indicated some lingering discomfort.

"Sure it can. The Chief designed it to take the change, didn't you!"

Hames rolled his blue eyes but was grinning uncontrollably in acknowledgement.

Emmalyne, in her usual way, cut to the heart of the matter. "So what do you two guys do anyway to make all this work?"

Chung's happy disposition faded. "Mostly, we have to maintain the repair engine interface." Chung pointed to an end point on the large grey block that still vibrated strongly in Hames' chest. Xavi moved to the end of the block and lifted up a panel to reveal an array of complex parts, wires and systems that appeared to glow and stretch downwards. On the far side wall, Chen Chung pressed his two hands to blank space, causing the wall panel to distort and reveal assorted tools and parts.

Emmalyne scrunched her face up as she came to a conclusion about her new crewmates. "So you're more mechanics than managers?"

"Yep, that's us, galactic handymen!" Laughing, Xavi and Chung put an arm around each other's shoulders and slapped hands together in a manner that made Hames feel embarrassed that this was his team.

The chief mission officer redirected the discussion. "Alright, kids, let's keep the tour moving. We've about sixteen minutes until the transition."

"Okay, let's take a closer look inside the block." said Xavi.

"The block?" asked Emmalyne.

Xavi ran his hands up and down the large grey shape in the centre of the room. "It's what we call the interface block. It's quite ominous to some people, the way it beats and glows. We try to put people at ease when they see it for the first time."

The four crew members began to walk a circle around the repair interface block as Chung explained its purpose, showing various parts in the open section and explaining the rest of the unit.

"You see? It glows beneath the multiple access points to show you where the functions are. Also, unlike other parts of the ship there's no molecule conversion on the panels that encase the interface block. We have to access the panels the old way as a molecule conversion process would be too dangerous this close to the repair engine."

Xavi took over. "The interface block lets us check how the repair engine is operating on a few different fronts. It's the main linkage point of the actual engine back into the ship's structure. For some fixes we can only access them from the outside, work on it externally from space. Hopefully we don't have to do that too much. The block here let's us change and manage various filters and conversion plates as much as we've needed to so far. The interface is also like a hub that directs us to the critical functions of the repair engine. It lets us monitor them, give them directions and review the matter being left behind in space."

"I've always found that sounded wrong, that we repair space by leaving something behind." Emmalyne was becoming more and more interested in the interface block and was peering into the backlit access panels.

"I guess you're right," said Xavi, standing still and contemplating this idea. "In any case, our process is simple really. We take in the damaged matter and reconfigure it into fresh black matter using a formula that incentivises the clean fragments to dissipate into nothing over time."

"Look here, this is the sample box." Chung placed his hand over a panel which slid down to show a round tube with intensely dark matter flowing into it and then pausing and swirling around. "We can hold a bunch of pre-converted matter for roughly a minute then let it flow through again; it's too unstable to hold on to for too long."

Emmalyne stared into the newly filled tube. Her features seemed to stretch towards it as if seeking to absorb its uniqueness. Drawn to another place, she appeared to Hames, for a moment, as somewhat ethereal. She spoke quietly to the substance in front of her. "It's beautiful. Like a bag of tiny black diamonds swirling all around each other."

"And dangerous!" Chung stressed with both hands closing the tube that held the black matter.

Emmalyne startled out of her thoughts. "And if the reduction rate is not significant you know something is wrong with the repair engine or one of its units?"

"That's right; you're pretty clued in, you know. It's a long trip. I think you're going to fit in just fine."

At this simple comment, Emmalyne started to breathe in quickly and sharply.

"Chief, my chest feels real heavy. I don't feel so good." Emmalyne suddenly dropped to her knees. Then she sat on her heels, her hands reaching for her sternum.

"Hey, are you okay?" Xavi moved to Emmalyne, who began breathing in faster and faster.

Between the breaths she tried to answer him. "I'm, ah, it's just a lot to take in." Tears welled in Emmalyne's eyes and then her hands went to cover her face.

Hames moved down beside her and reached over to hold her shoulders. "Breathe, just one breath at a time." Emmalyne responded and appeared to have more control over the moment that was taking her. Hames peered into her silent cocoon and spoke gently. "Let's try walking around for a bit." The two of them moved around the interface block a few times until Emmalyne was no longer out of breath or visibly upset. She stared vacantly now and seemed exhausted.

"I just need a minute." The young mission officer took herself to sit at the bottom of the stairs leading up to the upper gangway of Xavi and Chen Chung's workstation. Chung looked confused as he spoke quietly to Hames. "She seems real disconnected. Take her to Olisa, she'll fix her."

Anger swelled in Hames which he did his best to suppress. He too was breathing deeply, his strong shoulders heaving up and down. "Emmalyne's not broken; she's just feeling overwhelmed about the whole job. The whole thing!" Hames threw his hands in the air as he whispered assertively back at Chung. "She doesn't need fixing!"

Chung withdrew from Hames, returning to the back part of the interface block. "Okay, okay. I just hadn't seen this stuff before; people losing it takes me out of my mind."

Xavi moved closer to speak to Emmalyne, kneeling at her side. "Don't worry about it, we can handle it. You can handle it. Chief here tells me you're super smart." Xavi held his hands together and with a soft face consoled her. "Like we say in Universal Direction, we are here together. We are here and now."

Emmalyne looked up in recognition of this final statement. Some element of calm returned to her strained cheeks and tense hands.

"Okay, thanks. It's been different coming aboard here. Not what I expected to feel like."

"Don't worry about it," said Xavi. "The past is the past."

Emmalyne Biggs stood up tall and corrected wisps of blonde hair that had strained free from the tight ponytail she customarily wore. "Thank you, I feel better already."

Chung moved back around cautiously to help. "When I came on board I was a mess for a day or two until I saw Olisa. She helped me out no end; you should see her." Hames glared at him.

Colour began to return to Emmalyne's face. "I may just do that. I wasn't old enough to practise much Uni-D before I left home. Maybe now's a good time."

"Let's head upstairs and watch the screens for the change." Xavi offered a hand to Emmalyne and helped her stand tall, the four of them moving back up the stairs away from the interface block. "Even I get nervous too close to the block when we move in light alignment mode."

Hames burned at the suggestion that Emmalyne was disconnected, as if she was broken somehow. He hadn't considered how that might evolve to the prospect of her furthering a commitment to Universal Direction. After all, Emmalyne did seem to accept it. She would have grown up with it as a natural part of her life. The Flower of Connection was indeed the communicator symbol on her wristband. At this stage, she seemed far too young to be fully committed like most of the *Medens* crew. Xavi and Chung, though, were doing their best to entice her, despite their poor handling of Emmalyne's panic attack only moments ago.

Back at his workstation, Chung pressed a few buttons and the bench in front of him shifted into the shape of a small viewing tunnel, extending upwards from the bench top. He pressed his eyes into the darkness of the viewer. "It all looks fine in the matter scan. Just in time as we are about to enter light alignment mode in five, four, three, two and one."

The *Medens* started to emanate a round of slight vibrations. Emmalyne and Hames moved to stand behind Xavi who focussed intently on the workstation ahead of him. Hames felt the ship's vibrations begin to plateau to a feeling of normality as he saw their trajectory and velocity increase exponentially. He looked up at the dome shaped window above them and stars were now streaks of white. Finally, they were back at full repair capacity. Moving and repairing, ever closer to Mermai, yet still a lifetime away from where he needed to be.

Chapter 6

The Quadrangle

LYING IN THE STILL BATH waters of the cubicled washroom, Hames began to clean his greying hair. Even in the peace of moments like these, he felt the vibrations of the OUF *Medens*. Whether it was the quiet hum of power as the *Medens* floated slowly in the night or its unfathomable running with the flow of light, he was conscious of the feeling of frequency in his ship. This sensation was heightened during moments of stillness, his awareness bright and alert. Hames imagined the raw energy pouring into and out of the vessel as it churned through the particle debris in space. As the *Medens* progressed through space at the speed of light so did he.

The serenity was broken by an onslaught of latecomers. Hames could hear from outside his cubicle a selection of the crew enter the room. Mourning the loss of his solitude, he left his warm cocoon and began to dry himself. When he was done he pressed the two buttons on the top of the bath, the first sending the used water into the ship's recycling system and the second triggering the bath to reshape itself to the generic shape it had been before his arrival. Hames knew his cubicle, the first of four, would soon be in demand. Out of courtesy he began to dress himself.

To his right he heard a cubicle door open and a greeting come forth from Temett. "Morning all! First come, first served!" Down the row of cubicles Hames could hear the sound of other voices and activity. The perfunctory greetings exchanged between Emmalyne and Leena revealed not only their presence but the tension between them.

Hames was impressed that Emmalyne seemed more organised today; perhaps she was on the improve already. As the remaining steam and air dissipated from his cubicle he started to pull on his navy blue worksuit. He realised that today there was a chance they may both make their shift on time.

"Oh, come on, congestion in the hotwater system?" exasperated Leena. "Why is it me always waiting on the heating tubes. For mind's sake!"

Temett's deep and cheeky tone responded over the cubicle walls. "I told them to wait and save you a shower! Did that not happen?"

"Yeah, yeah, Second Pilot. Let's see how you go when I run you through the simulator this morning. You won't be joking when you revisit your breakfast in reverse."

"I retract, I retract!" pleaded Temett mockingly.

"Absolutely you do!" Leena laughed then switched track. "It would be great if the person in the cubicle next to me could turn down their hot water, maybe help the heating tubes reset?"

The atmosphere in the washroom instantly changed.

Emmalyne responded in satisfied tones. "Hmm, I don't think so. I was here first."

"That's not a very generous reaction," Leena mumbled, clearly caught offguard.

Water splashed as the precursor to Emmalyne's attack. "I don't know how you find offence if you Savai fanatics don't feel connected to anything."

The initial pause in response gave Hames hope that this childish argument had ceased, but it was broken by Emmalyne once more. "I think you must have a low level of inner self. Perhaps you're not very integrated."

"Why you! Erggh." Leena stopped at this point. Hames finished tying his brown shoes, readying to leave his cubicle, wishing that Emmalyne would soon do the same.

Silence persisted as showers stopped. Hames imagined frowning faces and hands gesturing through the walls of the cubicles as both protagonists readied themselves for the day. Leena's lack of an aggressive response did nothing to obscure the feeling of discord and Emmalyne set about restarting the argument. "Well, I didn't want to give you too much hot water. You might have washed that dirty golden spot right off your neck!"

Hames' mouth dropped open in shock at this full frontal assault. Leena could take no more. She left the shower, quickly wrapped her towel in a ball and threw it at Emmalyne as she too exited her washing space.

With a single hand Emmalyne caught and threw the loosely contained sheet back at her. Leena then hurled a small tin at Emmalyne, who ducked so that it hit the wall behind her, passing Hames and Temett as they left their cubicles.

"How dare you!" blared Emmalyne.

"How dare I, you, you, very rude person!" The inadequacy of Leena's retort was not lost on Temett and Hames and their laugher echoed around the wet and steamy environment.

With the last challenges still burning in their faces the two young officers dove forward for each other, both half dressed, Leena with her soft shells unbuttoned and Emmalyne with her work overalls hanging by her legs. Hames was not a man of action and intervention and he stepped back to allow space for the now dressed Temett to attempt division of Leena and Emmalyne, who rolled around the floor, pulling and cursing at each other. The taller Emmalyne seemed to have the upper hand, staying on top of her opponent. She moved to execute a slap on her smaller counterpart, but Temett caught her hand causing her to thrust her elbow backwards, catching Leena's protector in the head, hard. Immediately the fight stopped as Temett spun backwards and over onto his knees, catching the output from his gushing nose.

"Oh, Temett, I'm so sorry!" Emmalyne was now transformed, a gentle nurse patting Temett on the cheek as if this was the medically suggested cure for a broken nose. Temett tried to let out a word of reassurance to his assailant but all that came out was a guttural hiss as his own blood impeded his speech.

In the cessation of hostilities, Hames felt safe enough to assist. "I'll take it from here; you girls have done enough. Emmalyne, grab your stuff and I'll see you outside the washroom entrance point. Leena, you need to calm down; Emmalyne didn't mean it. You should get moving for your shift."

"Didn't mean it? Really!" Leena was indignant.

Temett looked up and around. "Is it over?"

"The mission or the fight?" Hames asked.

"It would be wishful thinking for the mission to be done. I would have to have been knocked out for a really long time." Temett smiled momentarily and then returned to catching the trail of blood from his face.

"You okay to go to work?"

"I'll be fine, Chief. I'll go see Olisa later; she can have a look at my face."

Hames said nothing and grabbed his things to go and catch Emmalyne. He was disappointed. This outburst made it clear to the rest of the *Medens* that she was still struggling. It was also likely to send out a message to Olisa to come and speak to this person who may need some of the tools of Universal Direction to calm her mind. He exited the washroom and found her not far down the corridors. Her face was drained and tired as she stared downwards to the conduits beating beneath them.

"You're going to have to live with that one. The event and that patch." Hames waved at the wet patches across her overalls.

"Yes, Chief," she responded despondently.

"Okay, let's get to our shift."

Time passed slowly for them both as neither was that keen on conversation and Emmalyne was worn out from her start to the day. Towards the end of the shift Hames decided to close out the day for Emmalyne. "You know you'll have to front up to the COO at some stage. She won't let it pass, and not because Leena is her cousin. You'll both be in the same amount of trouble."

"I know. I'm sorry. I just felt like making her feel, um, as crap as I do."

"That's not a very mature approach."

"I know. Oh mind. I keep saying 'I know' but I just don't have any control right now."

Hames looked at her and smiled. "You know, you can be a very rude person." They both paused and looked at each other, then burst out laughing.

Hames felt bad for sharing the funny side of Leena's frustration with Emmalyne, but his young colleague was doing it tough being new on the ship. "Why don't you take a breather and have an early break for the day."

"Thanks, Chief, you're awesome." She gave him a hug so tight around his neck that he felt his blood flow restricted.

"Sure, now get going." Hames waved her away.

The chief mission officer spent the last hour of the shift tidying up analysis for the next stages of repair; he was then ready to move his mind on to other things. Firstly, food. To be more specific, the food routine that he followed after all his day shifts. It was just unfortunate that Hames' routine happened to be part of other people's daily activities, or more aptly, rituals.

To eat in the mess hall, Hames would have to walk past the quadrangle. Its placement between the mess hall and the Recreation Room positioned it as a central location on the ship. This made it terribly difficult for Hames to avoid, although he shouldn't *want* to avoid it. Today as he arrived at its entrance point he couldn't deny that it was a beautiful place: white, gleaming and pristine. It was also the only open space on the *Medens* that you could feel like you were not jammed in like bio-circuits on a molecular communicator. Within the quadrangle was a large oval area, padded and comfortable to lie or sit on. Above this shape, a large oval viewing portal, the same size as the space below, provided a view to the stars and the path of the *Medens*. Despite its aesthetic beauty, the quadrangle made his skin crawl.

Most of the crew were gathered, ready to use the quadrangle for its designated purpose – to provide the crew of the *Medens* a place to practise the conventions of Universal Direction. Here they could meditate in close proximity to the Universe around them. For this, as Hames understood it, was one of the core beliefs of Universal Direction. The idea that all forms of life all came from the same matter, at some point in time, so are all connected. Hames struggled with the utility of this view. If he was related to everything and everybody then that made for one hell of an incestuous Universe. If his distant cousin was an atom or a rock on the surface of Pluto then he sure as hell didn't feel a need to visit. The practical application of this concept escaped him. However, after the Cataclysm the majority of humanity had adopted this idea. Most communities provided a place such as the quadrangle to supposedly enhance this connection and the *Medens* was no different. Tonight, the quadrangle was filled with *Medens* crew members all stretching and drinking water ready for their session.

Up the front of this group on a thick green mat was Olisa, the ship's healer. She called out to them all to find a free place on the mat so as not to intrude on each other's space. Hames found this terribly ironic. Olisa's white dreadlocked hair, with its matted patches, dropped around her head as she moved around the group, adjusting mats to ensure neat rows facing the front. The healer spotted him at the entrance point; again she invited him to join with her dry and cracked outstretched hands. He was supposedly a similar age to Olisa yet she appeared much older than him. Vainly he considered this a small victory.

Hames knew he did not have to comply with her unspoken invitation. There was no obligation for him to adopt this approach. There was no employment contract link. The only impost to Hames was that he had to live in a society where Olisa's approach to understanding the world around him was dominant. He had watched it grow from a simply political announcement just before the Cataclysm to something pervasive and all consuming; to fill the holes in people left aimless and wondering what came next. He had never been comfortable with it yet he had never been forced into it either. It was simply an irritant that he had to deal with on his daily walk past the quadrangle towards dinner in the mess hall.

As he'd done before many times, Hames shook his head at Olisa; a response unseen by the group in front of her. She adjusted a wide armless robe that covered her grey Ops uniform and seemed content. Although she had repeatedly asked him to join them she always looked relieved when he said no. Perhaps; this may have been Hames projecting his own relief onto her.

So it was that on most days he would walk past them holding their session. He could wait and go later but he had never been good at enduring hunger. In any case, he was curious to see what they were doing. Some days they were meditating and on others they were discussing the conventions of Universal Direction. Despite his disdain for the secular practice, he knew by heart most of these sayings, as well as the three pillars: meditation, curate-mente and liberty of consciousness.

Hames walked past the group and noticed Olisa taking the crew through a set of yoga-like stretches as she talked through what were apparently *challenges of unconscious thought*. Hames found that these were not actual challenges but impossible tasks. Just as he was nearly past and out of the uncomfortable sensation of the quadrangle, he found himself looking at the far back corner where he could see two bodies bent over, quite close to each other. As they moved upwards onto their knees he could see it was Leena and Emmalyne. They both looked content and to his shock they were holding hands. Hames was perplexed. This morning they were trying to throttle each other and now they were meditating together.

Hames' communicator began to buzz the tone of the chief operations officer. He moved away from the quadrangle and down the gangway towards the mess hall.

"Hames, I can see you lingering around the Quad. Come up to the bridge, we need to talk about Emmalyne."

"I'm a little hungry to be honest, can it wait?"

"It won't take long. Besides it will be easier to talk while everyone else is busy," replied the COO.

"On my way."

Hames made it to the bridge a few minutes later, the entrance point shifting in its molecules and structure to reveal its opening. Facing outwards beyond the clear glass he was struck by the whirl of space that blended in front of them. The lights of stars shooting past seemed to awaken the dark room. It also made the COO sitting at one of the desks look somewhat mysterious as the flecks of light ran across her face.

Hames leant his left arm over the back of Leena's seat. "Emmalyne, yes. I know you're concerned about her state of mind. Look, she's no different to others who have just come out of stasis. We need to support her, not push her into solutions that may not work for her."

"Okay, Chief, straight to it are you," the COO responded as her olive Mermain hands finished their manoeuvres on the desk in front of her. They both waited a moment until the COO sat back in her chair. "So what do you mean solutions that won't work for her? Seems like Universal Direction is helping her just fine right now." There was no sarcasm or smugness in the voice of the COO, just common sense and clarity as she pointed to the view screen showing the quadrangle.

"That's a solution, yes. It's not all simple meditation you know. There are the next steps, for going further in Uni-D. How do we know that's the right solution for her?"

"That's a reasonable point to raise. You're not her guardian though." Melini Guerra crossed her arms and leant over the screen to watch the quadrangle. "She seems to be pretty into it from what I can see."

Hames looked at the image. "They all seem pretty into it."

The COO had always offered some unity with Hames on his thoughts about Universal Direction. She had adopted it but retained an open mind to alternative opinions. "I see you on some days looking over the group when it's meditating or talking about things in the Universe. Sometimes you look like you want to join in."

The chief could not contain his sarcasm. "Clearly I've been holding myself back from this enjoyment all this time."

"You can't deny that the basic parts of Uni-D have value." The COO raised her arms above her and eased back in her chair. "There's nothing wrong with a good stretch and meditation. It's been a form of stress healing for thousands of years."

Hames held his belly with the broken compass right hand, trying to stop it growling. "You sound like you're reading from a brochure."

"Maybe. Space is stressful. So is wondering where we sit in the Universe." Melini Guerra looked out to the whirl of stars and darkness. "Especially now we've seen it and then been cut off from it."

Hames looked over some screens and absently checked readings of no consequence. "What about curate-mente? Do you think that is harmless? I know you've not taken it."

The COO pushed her long brown hair back, showing the flecks of grey that matched her age. "True. I'm not convinced, never have been. Others, though, have shown that it can have great benefit. Tell me, when was the last time you saw a crazy person in a transit lounge, talking away to themselves, or someone long lost to the colonies wondering around the streets homeless in body and mind?" Hames had no answer for this and continued to look interested in the stars ahead of him. She finished with, "There is without a doubt an absence of mental stress these days."

"There's also an absence in variance of character," Hames responded quickly. "We're starting to think the same. Emotions are not something to be suppressed."

"Maybe there's value in controlling them, not letting them dominate you."

"Emotions give us the colour and insight that can lead to inspiration, so can stress. If we don't have these inspirations maybe we'll never have the mind or imperative to solve the problems caused by particle debris and its destruction. It's been forty years and no-one has conquered it."

"Not even you, Chief."

"No." Hames paused. "The Cataclysm is part of history. It's almost starting to become long-term history. The current generation know nothing of the other ways. Uni-D was to be a pathway for harmony; perhaps it's not required any more."

"What do you mean?"

"It seems that the three pillars of Universal Direction are meant to be an ascending level of goals. Meditation leads to Curate-mente and that's supposed to lead to liberty consciousness. Lots of people claim that they've touched another level of life although no-one seems to be able to prove it."

"Isn't that the whole argument these days? Consciousness is subjective and experiential through one's own qualia?"

"Makes it damn hard to prove or disprove."

"Just because you can't prove it doesn't mean it can't be realised," suggested the COO.

"The inability of Uni-D to objectively show me that my mind works independently of the neurons in my brain tells me that perhaps it was never meant to, or, it just got lucky in that people needed it at the right time."

The COO made her own corrections in the screen in front of her. "Well, I never said it was perfect."

"Whatever the healing benefits it presents, Uni-D is fallible. Being told that a greater level of consciousness can be realised but you have to trust in the ideal...well, it makes it no better than religion." He paused to think. "It's just an approach based on faith."

"Perhaps." The COO seemed to consider this argument. "But forty years ago, like many other people, I needed Universal Direction in my life. There was a hole to fill and it helped fill it for me. It kept me alive." She looked at him directly, almost challenging him to debate the value that Uni-D had brought to her life. Hames thought better of it and decided it was time to discuss Emmalyne.

"Back to the girl. I'll keep a tighter watch on her, just for the time being."

"Okay, Chief, agreed. Now you'd better get some food into you before your stomach eats you alive." Melini Guerra moved around and back to watching over the functions of the *Medens*, signalling the end of the discussion.

By the time Hames left the bridge his stomach was violently struggling with his insides. He was starving and because of his conversation with the COO he would be delayed in resuming work. Work that he considerd more important than any mapping of repair routes or repair engine maintenance. The delay, though, had seemed to be the trend for the day. Emmalyne's antics had shocked him and then made him laugh. He had also spent hours preoccupied by her behaviour and what it may mean for her future. All up it had lost him time that was intended for his own nightly ritual of conventions and meditations: his search for Nisha.

Chapter 7

Clapham Junction

NISHA GAZED OBLIQUELY AT THE screen on her desk. The words showing the shuttle crew roster glowed intensely and blurred in and out of focus. Feeling blinded and inhibited she dug her fingers into her temples to alleviate the discomfort. After a moment she took a dark green pencil in her hand and began to draw. Slowly the shape of a sad mermaid surrounded by dark stones filled the pages in front of her. The image of Sovann Macha, a mermaid symbol from her home planet, came to life. Beneath this page the picture was consistently repeated, its colours and shapes reiterated by pencil over and over. These old tools of craft felt natural and comforting in her hand, sought after often during her times of stress or deep thought.

Hames had been right. Coming back to work was a bad idea. After all, it had only been four days since the explosion outside St Paul's. She had thought she was fine after getting over the initial shock of the incident. Hames had stayed with her for a few days and by this morning she deemed it unnecessary, sending him back to study at the Agency. Again she found herself looking out at the barren rooftops of south London. It felt a waste to be there alone, realising how lucky she was. When she first arrived back home many tears had been shed at the realisation that others had not been so fortunate. In time, though, she felt more on top of herself and obliged to get back into life.

Today, just as she returned to work, the physical pain started to set in. Her lower back ached, the left shoulder was intolerable and her right wrist started to emit short sharp bursts of electricity at certain movements. Nisha had mentioned the wrist to the doctor at the Charing Cross Hospital, once she had finally been seen. Why she was sent so far out of inner London to Hammersmith she did not know. Her memory of the post-blast events was vague. The explosion at the Universal Direction meeting had thrown her metres into the air, thudding her side-on into a bus shelter. Luckily, an older man below her had cushioned her final fall to the ground, although she had strained the muscles in her lower back. *Damn old man, I might have just landed on my arse instead,* she thought to herself.

The Academy and Hames had a connection to Charing Cross Hospital, something to do with the new Med-Clinic technology that could patch up pulled and torn muscles in hours. They tried this on her shoulder and lower back, and it seemed to help at first. Now it hurt so bad she wasn't sure if the new technology had actually been switched on. Her recollection was quite hazy, which was surprising as she had not landed on her head so hard that it should cause this vagueness. She had only received some slight bruising to her face from limbs moving around her in an uncontrolled fashion, possibly her own.

What had happened was intense. The explosion had felt like a wave of power throughout her body. An explosion in which she had seen others close to the detonation point pulled apart violently into nothingness. Just dust and debris where they used to exist. This experience was followed by chaos. First, screaming and shouting, and then begging by maimed attendees of the peaceful public meeting. Sirens and more shouting had silenced the bells ringing in her ears, announcing the arrival of military and medical teams. There was almost a carnival type atmosphere with a full fanfare parade of uniforms, megaphones and lights that penetrated the twilight of the day. During all this commotion the unwelcome heat had put a layer of further oppression on the events. It was a level of sensory overload she had never experienced before and it had taken her three days to get back a sense of normality in her perception of the world.

Hames had been fantastic, rushing to the hospital and making sure the right doctors and nurses saw her. After a few hours it was clear she could move about with relative ease and it was just bumps, bruises and strains that she would have to deal with. All in all, her physical state was acceptable and Hames was right, she did look tough with a bruise running down her face. He had wanted her to stay home and stay safe. As if she was the sole objective for the militia that had targeted the Agency's new program of healing and harmony.

As for Jasiah; well, what a lucky bugger. He had not been thrown through the air as she had. Rather, he had been caught in a throng of people who had been squished together then toppled backwards over each other like a stack of dominos. He had called in to see her once on the first day after the explosion, complaining about how terribly winded he had been from all those people half-lying on top of him. Hames and she had waited expectedly to see Jasiah again on the second and third days. Prepared picnic lunches for an excursion to the local park to enjoy the midday sun went unattended by her cousin. They assumed that despite a lack of injury, Jasiah also needed time to heal. At some stage Jasiah would have to come out and say hello, or at least provide some signal to show he was still around.

So here she was at work. A bad decision, she thought, as she continually rubbed her temple, trying to focus. Her boss, Peder Jonslow, came into her office and his bulk curved over her desk as if he was about to share a deep dark secret. "Nisha, there is someone here to see you. A man from the Agency. It's about the, um, event the other day." Despite his firm character, Peder was clearly bothered by being so closely linked to such a dramatic issue.

Nisha was relieved at the expected distraction. "Oh, thanks. Hames said this would happen. I'm happy to speak to him, where is he?"

"He's in the Eastbank meeting room. Did you want me or someone else to come with you? You don't look well."

Peder looked relieved as Nisha shook her head. "I'm okay. If I start to feel rubbish I'll excuse myself."

"That's fine, take your time. I wasn't expecting you back today or the rest of the week!" The large blond man gestured to the door. "You can take the leave that is required."

Nisha stood up quickly and felt a dizziness that almost brought her back down. "I think that's a good idea." As she walked down a long corridor towards the impromptu meeting while she began a message to Hames:

> Just stepping away from my desk to meet a man from the Agency. You said someone would come to ask questions. Anything I should know about the Agency Security Stream? Miss you, I should have stayed at home...head hurts...

After more strides, turns and the casual greeting of a few people in the walkway, a message came back from Hames. Normally he didn't reply to her until much later. Today, he must be checking messages, maybe even waiting to hear from her. She quite liked this change in attention level. It made her feel special.

Sorry your head still bad, you should be at home!! Agency security are no worry, but they do monitor aliens to Earth like you! Tis just protocol, nothing to worry about. Am thinking of you non-stop today. You're my everything, can't lose you, would break me.

Nisha felt tears come into her eyes. She was now at the meeting room so it was time to focus, with time for one quick reply:

You'll never lose me! My heart is like a compass and it always points to you, draws me to you. Never lose me, never.

After this heart-felt message she looked around the corridor to see she was alone. *Thank the stars* she thought as she wiped her wet and sniffling face down the arm of her blue work shirt. She slowly gathered the cuff into her fingers and dabbed her eyes dry. Feeling more presentable she placed her hand on the door panel and entered the meeting room.

A small man in a tight charcoal suit looked up to greet her. His size and calmly extended hand made Nisha uncertain as to whether she was in the correct room.

"Hello, I'm Sean Watson from the Agency." A glimpse of a smile added to the short brown hair and freckles did little to ease her uncertainty. The man was not what she was expecting. Diminutive and open in his body language, he had the hallmarks of average all over him.

"Hello, I'm Nisha Pinto, but you already know that."

"Yes. Yes, of course." The man sat down at a nearby table in the windowless room and Nisha matched this action. She was struggling to get a feel for who this person was and what he wanted.

Sean Watson settled himself and loosened his tight jacket. "How are you fairing since the incident the other day?"

"I'm fine, thank you. Alive, for which I am grateful. That doesn't stop me feeling bruised and sore all over."

The man from the Agency offered a sincere nod of his head. "It was a terrible tragedy." Sean Watson paused. "Nisha, can I ask how much you are aware of the Eagleton Accord?"

Nisha recalled her discussion with the blue-suited woman at St Paul's, just before the explosion. The woman in front of her had given a strong decry of the Accord and praised the creation of Universal Direction. "Well, I was very aware of it before the explosion. Since then, even more."

The man ahead of her shifted in his seat. "That's an interesting way of looking at it."

Nisha still felt passionate about her home and was not going to hold back, regardless of the Agency or unexpected explosions. "You know I'm from Mermai. So to me the Eagleton Accord represents what happened when the outer colonies began to establish their own identity. It was when we understood our future was ours, not someone else's." The man looked back nodding with her statement. For the briefest second she felt calm, that he understood what she was saying. However, something told her otherwise. She didn't know what it was, but she felt there was another matter at hand. This unease, though, didn't quell her passion for colony independence. "I appreciate some people on Earth felt a loss when the colonies seceded. In reality, it's just not damn practical being twenty-odd light years away from the central government. Particle drive travel or not."

Sean Watson opened his palms on the table. "I can understand that point of view. In just a hundred years the colonies had been able to ascertain what were the challenges for their planet and what was needed from their environment in terms of resources for future generations to survive. You have to protect the natural state of the ecosystems as much as possible."

Nisha cautiously agreed. "Okay, go on."

"From the early days of settlement, on all colonies, not just yours, there grew a strong desire for their planets to avoid Earth's own challenges. To avoid congestion and an overused or bastardised ecosystem. Eight years ago, when the colonists came together to lobby the Interstellar Agency for a restriction on any further mass migration and to seek self-determination, while it was resolved peacefully it wasn't exactly what Earth wanted. It wasn't what the people of Earth wanted."

Nisha cut in with her hand chopping down on the table. "People tend to forget that the Agency agreed to the proposition from the colonists. They agreed to self-determination based on contracts for continued trade and access to ports. That is what Nathaniel Eagleton agreed to for the Agency." She held up her hands in exasperation. "The colonies got their independence and the Agency got its trade routes. It was a fair deal was it not?"

"Yes, yes it was. But with that peace came monitoring. You may not remember, but there were some testing flashpoints of threats and acrimony in the lead-up to the Accord. Many people on Earth didn't understand why the colonies wanted to secede. The Agency did and could see the benefits for all." Sean Watson hesitated and tilted his head sideways. "It just did a bad job of explaining them."

Nisha rolled her eyes. "I can't argue with that. Some people on Earth still look at me as if they own me."

Sean Watson began to talk more excitedly. "That is one of the key reasons for Universal Direction. To show connection is present across the worlds without the physical ownership by Earth."

"Where are you going with this?"

Sean Watson leaned forward, his plain, freckled face energised. "Since the Accord there's been a constant watch for a level of aggression against those on Earth who opposed the loss control. As you know, some people hate the Accord with vigour as it represented a step towards a galactic identity. An idea that also upset people on the colonies. There are lots of people out there simply opposed to the idea of harmony between Earth and the colonies."

Nisha was tired, sore and becoming exasperated with the discussion. "What has all this got to do with me?"

"Well, when we analysed the event at St Paul's, looking for who was there, you and Jasiah were identified as colonist born. While I certainly don't believe you had any role in the event I have to tell you that you're officially under investigation and protection."

"What!" shouted Nisha. "That is ridiculous! How can you think it was me? I wasn't even meant to be there!"

Sean Watson sat back upright in his seat offering reassurance. "Please calm down, it's just protocol. You were there and based on our policies we must follow the simple leads and indications. You're from a colony so the action flows from there."

Nisha was almost laughing at all this. "You people from Earth always feel like you're owed something. We're not cattle, though, we're humans too."

Sean Watson seemed satisfied with himself, as if he had achieved the purpose for his meeting with her. "Of course you are. I don't even need to say that."

Nisha's mind then turned to a previous statement and a word she hadn't quite registered at first. "What do you mean protection, as well as investigation?" Her emphasis placed on the former word.

"In practice, we have reviewed your case and our computers have analysed your potential for two critical outcomes. You could be a person who may attack others, hence the investigation." Sean Watson looked apologetic at this statement. "Additionally, you have been identified as a person who may, in fact, be attacked themselves."

"That doesn't make any sense. How can I be both?" Nisha's headache was getting worse. She was completely over the interview. "It's ridiculous."

Sean Watson himself was puzzled at this incongruous outcome. "It's correct that you are the partner of one of the prize students at the Academy, if not the most important student and staff member of the present day. Much has been put into Hames Naughton and his ability to map routes to new exo-planets."

"I guess that is true." Nisha had witnessed first hand the lengths people went to in order to keep Hames happy and working hard. They had been taken care of exceptionally well by the Agency.

Sean Watson continued. "Given your special linkage to Agency personnel you've been flagged by our system as requiring monitoring. We would appreciate your assistance with this; it does make it far easier on all involved."

Nisha thought of Hames. She didn't want to spoil things for him. She believed he loved his work as much as he loved her. In any case, there were only four weeks to go on Earth. She slumped back in her chair. "Fine, fine. Go ahead. Monitor me all you like."

"It would only be for a short time, days, in fact, I expect. We really don't think you're any kind of suspect. We just want to keep an eye on you, for your own protection."

"What about Hames, does he know?"

The small freckled man from the Agency suddenly looked more diminutive. "No. We were kind of hoping you wouldn't tell him, that he doesn't need to know."

"Oh, you have got to be joking!" The shake of the Agency man's head told her there was no joke. Once more she begrudgingly agreed. "Okay."

"Thank you. I'm sending you my details now. If you have any further questions or updates on your situation please let me know." Watson pressed a small square communicator screen that he produced from his pocket. "When we find out who was responsible we'll let you know your situation has been resolved."

Nisha looked him directly in the eye. "Let's be clear. I don't have a situation, you have a situation."

"Yes, I know. My apologies."

With her final curt statement delivered, Nisha stood up to leave. "Goodbye, Mr Watson." She made the long walk back to her office-pod, each step along the corridor a further push away from a willingness to work. Passing under the large screen that loomed over them all, the movements of shuttles up to the particle conversion ships gave her a reminder of one last task before she could leave for the day. A rare task and one that she would not allocate to others. She felt Peder's thudding steps follow her into the office.

"Nisha, you do not have to meet the delegation from Sol-Concio; you can go home. They have their own shuttle platform to arrive at as well as their own shuttle pilot to take them up top."

"Yes, but they're the only group to have a free pass through this entire complex. I don't like it. We're supposed to clear anyone moving up to the platform!"

Peder rolled his eyes at the old argument. "They don't need us! It's a cleared diplomatic path from London to here then up to their particle ship. No-one else needs to go near their departure point!"

"Don't you want to make sure nothing incorrect is happening in your complex?"

"It's not our issue! You're just too curious, Nisha!"

"Perhaps I am. Look, they may have their own diplomatically approved path up to space. Whatever. There's nothing stopping me from checking our exit and entry points, a safety inspection if you like." Nisha grabbed her jacket and left for the arrival point at the far tower of the shuttle docks. She felt improved after her discussion with Sean Watson, almost invigorated.

As she waited to meet the delegation returning to Sol-Concio, Nisha looked across at waves of the Channel, which grew in venom as they slapped each other harder and higher, spraying mist back towards the metal landing bay. A small warning alarm signalled the arrival of a golden transport craft that surged into the bay, cowering the seas below in its arrival. Two men and women departed the ship, hauling their own boxes and bags of what Nisha could only think were supplies of some sort. Their loose yellow and white layered work clothes allowed ease of movement. They all seemed old, although, their deep tan skin gave them a youthful glow and their movements were strong and certain. To Nisha, they all looked very similar.

One of the women with honey-brown hair flowing loose in open long curls approached her. The delegate to Sol-Concio extended her hand. "We appreciate your time here, Officer Pinto. I do remember you checking up on us a few months ago. As then, as now, we are fine thank you."

Nisha circled about the area as if she had some form of actual authority in play. "What do you have in those boxes?"

"Just some earth based minerals for building materials. The rest we don't care to divulge nor are we required to, you know that." The woman from Sol-Concio sounded polite and Nisha detected no threat in the delegate's body language. The woman was confident that the boxes they were carrying would be going onwards unexamined. This irked Nisha somewhat. She had never learnt much about Sol-Concio. Then again, no-one across the colonies had an answer about why Sol-Concio was treated so differently. Why it was treated in such a special manner.

"Why do you need a separate route from the other colonies, your own departure point?" Nisha asked.

The representative from Sol-Concio looked around at her colleagues and encouraged them to collect their cargo. She turned back to Nisha as she started to leave. "Good day, Ms Pinto, we'll be taking the lift by ourselves." The four members of the delegation piled into the lift with their boxes, leaving Nisha alone, listening to the waves battering and breaking each other endlessly.

In London that afternoon, the grey clouds that punished the Dover seas now covered the sky in dark grey tones. As Nisha moved through the Clapham transport hub the lighting strips on the rusty red-brick buildings around her switched on in recognition of the darkening skies. The warning on her communicator suggested rain was soon to arrive. "Perfect, pissed on from top and below," she said to herself as she walked slowly along the street.

Nisha was under watch. She had no idea where from, she just knew someone was there. *I bloody better buy some tissues* she considered. There could be no more wiping her nose with her sleeve or scratching her bottom. All her private moments would be shared with others. How infuriating this would be. She pulled her communicator out from her pocket to check the rain warning and saw the delivered contact details of Mr Sean Watson waiting for her. She reluctantly pressed the confirm button and the details vanished into the memory banks, forever etching the scrutinised situation she now found herself in.

Her right wrist began to ache intensely. She had received all the allowed painkillers so it was now just a matter of grinning and bearing it. Creature comforts from home were required and the woman from Sol-Concio had sown an unrelenting seed in Nisha's mind: chocolate. A short walk to 3 Lavender Street brought her to the Memories of Mermai. Outside, a rich grass green–painted frontage and a wide window displayed dark and light brown delicacies in the form of pastries, breads and meat. Surrounding these offerings were tiny bright blue packages of various shapes indicating the sweet offerings from Nisha's home planet.

The welcome faces of the shop owners, Rema and Nash, made her feel comforted any time she walked through the door. After explaining her battered state to her friends it was time to select her treats from home. Big bars wrapped in the bright blue paper contained the smoothest tasting chocolate known alive, made with the milk of the Mermain cow. She also bought a kilogram of various types of beef spiced with the flavours from the rich dense forests of Mermai, and flat breads from Mermai-Goa, as aerated as normal bread and with a taste so distinct the strongest herbs could not overpower it.

She was equipped and ready to lie in state for the rest of the week. Nisha departed to reacquaint herself with the greying afternoon on her walk back into Clapham. In front of the station a young couple handed out paper pamphlets. It was an old-fashioned practice and caught her interest. The young man reached out to her with a large smile and offered one to her. "With light," was his greeting.

Nisha accepted this and placed it in her pocket. As she waited for her train she retrieved the paper. She saw no words, only the image of a sun. She touched the picture and it transitioned to the eight-circle image of Universal Direction, the one shown on the banners outside St Paul's. The image spun until she touched it again to reveal a message:

Universal Direction – the time is now!

Ealing Court Stadium this Friday to hear Legeles de Nord outline the pathway to a new way of living together. Through Meditation, Curate-mente and a Liberty of Consciousness we can have a better place and time.

Four conventions to guide you on this path:

The past is past
We are here and now
What will be, will be
You are the everything

With these steps Earth can once again be harmonious and move as one with the stars.

Nisha strained to read any more of the words that began to scroll down. Her head hurt too much. The train arrived effortlessly and she took her seat. The ideas from Universal Direction – there seemed some need for them. But did people have such a gaping hole in their hearts that needed to be filled? She felt only love in hers, especially for the bag of comfort food that lay at her feet.

The train pulled up and she started the short walk home. Her body was still sore and the remedies from Memories of Mermai, while welcome and necessary, would not be enough. She needed to see a doctor again. Nisha cradelled the communicator in her claw-shaped sore hand as she typed 'doctor' 'battersea park'. Instead of the usual options coming up there was a new title now showing 'Doctors & Healers'. A new practitioner nearby her apartment block in Battersea Park Road looked free. It was closer than the one she had been seeing in Clapham and it seemed reasonable to book an appointment for the next morning. What was the worst that could happen in trying a new doctor? Especially with her own security brigade watching her every movement, wherever she would go.

Chapter 8

View of Mermai

THE MESS HALL ARRANGEMENT WAS as Hames expected – reformed and reset with multiple smaller tables rather than the large single shape used for team meetings and planning. His hastily taken meal in a far corner consisted of roasted sweet potato, beans and miso. Energy for the night ahead. This relief was broken in the shock of Emmalyne planting herself without notice at his side.

"Chief!"

Hames jumped. "Damn it, Emmalyne! You scared the life out of me!"

She gave him a smug wink. "I know."

A hiccup came from Hames, affirming the physical reaction that his colleague had just generated inside him. His attempt to scowl at her was instead replaced with a grin. "See what you did? I'm bloody hiccupping now. Haven't done that in years."

"Suits you, Chief. Hey, am I in trouble? I mean with the COO?"

"No, no, she just wants us to make sure we all work hard to get along with each other. It's not always easy on these trips."

"I think Leena and I will be fine," Emmalyne said. "She's quite funny in her own way. She's snooty and has strange beliefs, but she cares about others."

Hames hiccupped at her again. "That's one way of putting it."

"Tell me more about Leena and the COO. You said they're cousins; how did they end up here?"

"There's no secret about Melini Guerra's history. Although it's not something she likes to extensively share." Hames pushed back from the table, "Wait a moment, I need a coffee to stop these hiccups." Hames got up and started programming the dispenser. A small wall panel shifted and moulded to reveal an opening, allowing him to grab a plastic cup with steaming hot black coffee inside it. The smell was already invigorating him.

"Chief, you know what the Organisation says about coffee."

"That it's a neurotoxin slowly poisoning me to death? Yep, but I'm working late tonight and hot chocolate makes me relax, not work." He hiccupped once more and squinted at his young workmate.

Emmalyne jumped up, clearly in a playful mood and joined Hames by the dispenser wall. Inserting her hand to the panel, the wall opened again to review a hot pot of chicken soup.

Hames shook his head. "You know it's not chicken, only a substitute."

"The same goes for your coffee! Do you think there is a secret stash of beans in the wall somewhere?"

He played with his beard thoughtfully while taking his seat again. "Now that is a nice idea."

"In any case, we don't eat chickens do we? It's not very vegan is it?"

"Perhaps not. So you know the COO is from Mermai right?"

"Yep, it's pretty hard to miss."

"And you know Mermai was settled out of Indian migration, southern India mainly. Not long after the discovery of particle conversion and new colonisation tech, Indians were one of the first offered the chance to start afresh."

Emmalyne blew on her soup, her high cheeks becoming rosier then normal. "My early 23rd century history is lacking. Why was that?"

"Overcrowding and a whole other range of issues." Hames took a sip from his perfect temperature coffee, its bitter warmth filled him contentedly. "So when the opportunity came for a group of them to start again in what would eventually become Mermai, tens of thousands took up the offer. It was the most subscribed of all the movements. Mermai is a place that typically doesn't have much weather of note. It's a calm planet and its peacefulness was like a metaphor for what these people needed."

"How do you mean no weather?"

"It only snows and rains in the mountains, nowhere else. There is very little in the way of wind and the movement of the planet is such that on its equator the temperature is highly consistent."

As her food cooled, Emmalyne started to take in spoonfuls of pale yellow soup. Hames could smell the heartiness over his coffee. She stirred the bowl while looking for answers. "Okay, I get that the COO is from Mermai but Leena wasn't born there?"

"Right. The COO is a multi-generation Mermain. Enough to soak up the atmosphere in skin, hair and body. Even though she still had a big family back on Earth. She was in her late twenties when a message came through about her grandma, who had moved back to Chennai. The grandma was ill so Melini decided to take a trip back to Earth to see her for the last time. The COO was on Earth just a few days before the Cataclysm occurred and she was stranded."

"That's horrible," whispered Emmalyne.

"Like I mentioned the other week, I'm not alone in being cut off from loved ones. Thousands of us went through this."

"Oh, Chief!"

"It's okay, really." He continued his story. "When it all went down there was a big effort from all sorts of groups and bodies to come together and build a support service for the people cut off from their loved ones. The Interstellar Agency, which I worked for back then, had to transform from a body that coordinated the colonies across the Universe to a new type of 'organisation'." Hames stressed the final word in both his voice and expression.

"I know, that's where the Organisation came from." Emmalyne leant back in her seat taking a moment from her hot pot, which was still dispensing whisps of steam floating into the air. "The Organisation seems so big and bureaucratic and, unhelpful." Surprise came over her face. "It's hard to imagine it being set up in response to what was a humanitarian effort."

"But it was. Over a hundred thousand people were stranded between the worlds. And not just people were affected – there were thousands of merchant entities that over decades had established themselves based on trade from the outer colonies that closed overnight. The Organisation quickly became the defining body for our society; it started to run everything from food supplies to power provisions. It also pushed Universal Direction out across the community."

Emmalyne redirected him. "Again, back to the COO and Leena."

"Sure. As you know, they're related. It's all tied up with the COO's dislocation. Leena herself is not dislocated. She's from Earth and she's pretty much the only family the COO has. As painful as the Cataclysm was,

I was lucky. I lost just one person, the COO lost all of them. Her son, her daughter, her husband, her mum, her dad, her friends, who knows, even a pet fish or dog.

Hames wondered if this was finally enough. Emmalyne was looking glassy-eyed and stiff-faced. Her body no longer excited, her soup unattended to. A small tear escaped her left eye and she started to breathe deeply. He could see her processing the story over and over in her mind, going deeper and deeper, as she picked through the realisation of what had happened. It was one thing to read the stories of people being dislocated, another to hear about someone she had met in person, the leader of her ship. He had let her keep going when most normal human beings would have attempted some form of consoling. He had wanted her to understand the story of the COO, but on reflection he realised it was a selfish act, just as much to reveal his own pain. He clenched his fist and pounded it silently on his leg.

Emmalyne began to control her emotions. "What happened to the COO's grandma, did she get to see her?"

"She did, and the grandmother got better and lived another ten years, even past her hundredth birthday."

After the mess hall, Hames returned to his pod for his reports and calculations, the work he hoped would one day provide him with the answers he yearned for. His pace through the winding corridors of the *Medens* quickened to compensate for the delays of the day. A short time later he was back against the back wall of the dark-grey Recreation Room that lay between the quadrangle and home-pods, reading a tablet on old PCD routes to Mermai. He was waiting patiently to use the large sphere-shaped mechanism that Leena was programming. She had been ahead of him by just a few moments and she appeared to have plans to use the ten-metre wide viewer for at least some part of the evening.

Leena stepped down off the platform that surrounded the spheric device. "Don't worry, Chief, I won't be long with the Williams Sphere. I just wanted to have a look at Mermai with Xavi."

Hames tried his best to hide his frustration. "Not a problem, take your time." He had wanted the sphere to walk the space route to Mermai. To walk as if he were part of the stars himself, alone in the black and damage. It would shift and mould around his steps, letting him traverse the route a particle ship would have travelled many years ago. In reality, though, the Williams Sphere would swirl in the same spot making the image of movement throughout space seem real.

Leena stood next to him looking over his shoulder. "Working alone again tonight?"

"You know how it is, out here you lose track of time and people."

"Chief, you could take a lesson or two from Uni-D. I know it's not your flavour, some parts of it aren't mine either. Working harmoniously together sounds pretty good to me. Sharing is part of that, yes?"

Hames looked up from his tablet, his mind and blue eyes lagging behind in their response to Leena. "I'm sorry?"

"The sphere, I know you were looking to use it. You do seem to have it every second night or so. I think tonight it's okay you share it for a bit?"

"Of course, of course. You're right. I do seem to make use of it more than others. I reckoned that was down to the fact that most of the crew can't operate it."

"I can work it. Don't you worry about that." Leena gave him an assuring wink. "I've come here a few times to use it just to see you walking intensely across a space map."

Hames gave a half-grimace in acknowledgment of unintended shared moments of investigation.

Leena distractedly brushed her finger over her wide gold sensor spot. She looked lost in her own thoughts for a moment. "I know why you come here; you're searching for something you lost in the Cataclysm."

"Maybe, I guess it's my time. There's no issue with looking back to see why things happened, is there?"

"You don't have to look back, Chief. Universal Direction shows we are all connected. The way my Savai branch understands this connection is that being linked doesn't mean a loss of self. We're not all going to turn into non-corporeal beings and float away all together in a big pile of gloop."

Hames looked at her with absent and quizzical eyes. "That sounds horrible. I'm glad it's not the case, as far as we know."

"It's what we say it is. Consciousness is in superposition until you decide where you're going."

The way Leena expressed Universal Direction always made it sound a palatable option to him. It *sounded* far more open than the actual practice. Mainly, though, Universal Direction felt to him like an excuse for disconnection. If he stopped looking for Nisha, even decades later, he would feel a failure.

Xavi entered the recreation room clapping his hands together in readiness. Leena concluded her lecture. "I'm going to take Xavi down the rapids of the Bullet River. He reckons he'll cope...hmm, I doubt it!"

"I've been down the real ones. I don't think I used the paddle once. Just a lot of mindless screaming."

Leena's shoulders relaxed. "I remember you telling that story last year. It's stuck in my head ever since. Xavi's too. Maybe one day you and I can try them together." After starting to walk away Leena paused. "You don't always have to pine for the past. There's a lot of individuals here and now, happy to hang out with you."

Hames did his best to take this positively. "Okay, thanks, Leena."

For a long time he had been searching for evidence and clues as to what had happened to Nisha. He had lost her on the day of the Cataclysm, gone except for a few breadcrumbs to follow. Over the years he had excluded many possibilities. Now, he believed her disappearance had something to do with the Mermai Tentacle. He began following an idea that he could break down the PCD debris trail that formed the tentacles of damaged space. This might show which ships had travelled this way before. In theory, he could analyse the debris to identify different types of engines and then the actual ship. With enough data he could look at the age of this debris trail; that would be a critical factor to determining if a ship Nisha was on the day of the Cataclysm had made it along this route.

There was small hope that the effort of investigation tonight would yield a conclusive answer. It was immensely complex and detailed work, a long-term project. Now it was more just a habit to keep looking. He would search for Nisha until he could look no longer. Raising his aged head up from his calculations he saw Leena finish her entry into the screen next to the sphere. A space in the large plastic-like ball began to distend, allowing Xavi and her to enter the sphere. Outside, viewing screens showed the two new entrants walking into the object and stopping at its centre-bottom. It was like they were inside the Earth looking up at the northern hemisphere.

Leena had linked the leather band communicator on her wrist with the Williams Sphere. The symbol of Universal Direction glowed as she activated the device. "Let's watch some analysis of Mermai first, get some detail."

"Sure, whatever you say, you're in charge." Xavi gestured casually to the faded images around the inside of the sphere from its previous point of use. "You'll have to delete the last file, looks like someone was on Cernunnos."

Hames put his work down and watched the viewing screen that showed the interactions in the sphere. Leena tapped on her communicator and a picture zeroed in on the *Medens* and where it was now in space. This picture then zoomed out; as it did, a counter of the distance from the *Medens* began to tick over.

"Hey, you know what I heard the other day in the mess?" asked Xavi as stars and constellations sped around them.

"No, I don't. You will tell me anyway," said Leena with a raised eyebrow.

The picture zoomed out far and fast from the *Medens* to show a large portion of the Milky Way.

"Francois was in the Cernunnos-Lugus Civil War, apparently he fought in the Battle of Caltrap."

Leena gasped. "No! I'm obsessed with that battle. The footage...tragic and amazing!"

"Yep, I know you are. You have to ask him about it."

Finally, the picture around them shifted to a point twenty light years from their current location. This view started to focus on what was a very intensely blue and green planet. Hames listened to the viewer and watched the screen displaying the contents of the sphere:

> Mermai, previously known as Gliese_667_Cc is twenty-three light years from Earth, making it one of the closer outer colonies when compared to Sol-Concio and Cernunnos, which are thousands of light years from Earth. From the early 2100s, images and data produced great confidence in the capacity for habitation to occur at some point on Gliese_667_Cc. It was the second planet to be settled through the use of travel by particle conversion drive technology following Earth-2. Mermai was initially colonised in 2310 and was renamed by the young astrophysicist who planned the initial particle drive route, Bala Garang. When Garang first saw pictures of the northern hemisphere it

revealed a large ring of islands above a fish-shaped continent that resembled a mermaid.

As the sphere detailed this location it zoomed in closer on the area around Leena and Xavi, beginning to look, feel and sound real. The sphere reimaged itself from its point in space down to the planet below; a long green hill that was backed by low-lying rocky mountains. It was a similar view to Earth except for the aqua hues that reached out from the long swaying grass and the mountains in the distance. The whole visage of Mermai seemed to glow in a cerulean blue.

Leena and Xavi stood atop the hill, looking down at the prairie then gazing up at the mountains. Wind blew softly and intermittently through their hair, freshening their senses. They both looked calm and relaxed, almost like they were in a trance. The two crewmates started to walk across the crest of the hill, which wasn't high but stretched out for over 200 metres. Towards the end point of the rise they could see the hill progress down to its base where there were fields to the right and a forest to the left.

Leena pointed down to the fields. "Let's go over there." As they moved forwards they walked very carefully. Leena and Xavi were not actually walking on solid ground; they trekked inside a shifting and morphing metallic ball. It was like a giant grey jellyfish had swallowed them whole. Their movements in this beast could appear particularly strange and awkward. As they strode across the crest of the hill, their creeping style of walking made them appear to be hiking on fragile ground. As best they could, each matched their footsteps to the other.

The strange imagery of his two colleagues walking so distinctly across a hill in Mermai was no distraction to Hames from his work. He glanced to his right to see the large sphere moving at a steady pace, rolling forwards over and over but never moving anywhere. Xavi and Leena were not doing too badly in managing the strange sensations the sphere could impose on a person. Hames had watched some people get totally disorientated, trip and hurt themselves. Those who took training and were proficient in its application could gain great benefit from its use and Leena was well practised in use of the sphere.

Xavi looked confused as they came to the end of the crest. "Why does the grass look green up close but from a distance it has that kind of blue aura?"

Leena looked curious. "Not sure. Let's listen and hear what it says."

Although a super-planet, much larger than Earth, the environment and ecosystem were ideally suited to human population and bore many similarities to our home planet. The topography of Mermai is such that blue elements of...

"Hello, Chief, how are you this evening?"

At the entrance point to the recreation room was the ship's healer, Olisa. Her hands were crossed and she was smiling in an intense manner. An intensity that was reinforced by skin that was weathered and craggy. Olisa was also thin to her bones and this combined with her drawn appearance made her a stark figure. She always seemed to hold her body so tightly it made Hames feel instantly on notice for some kind of question or situation of discomfort.

With a hidden sigh Hames wished that Olisa hadn't appeared in the recreation room. He normally tried to avoid her company, especially late at night when the ship was quieter and there was a greater opportunity for earnest conversation. The plain truth was he never accepted or engaged with the brand of healing that was the domain of practioners such as Olisa. Med-bars provided enough maintenance of his own body and after that, in his view, there was nothing left to be done. He tended to disengage from her as he saw nothing overtly wrong with his mind, nothing that required Olisa's help or correcting.

The healer waited in the doorway. He considered saying something uncomfortable yet truthful such as 'how convenient for you to find me here' or 'I really don't want to talk to you right now'. However, he knew that this was an impolite attitude and there are some conversations that are ultimately unavoidable.

"Hey there, Olisa, come on over." Hames offered the healer a seat next to him in front of the screen showing Leena and Xavi's adventures on Mermai. By this stage Xavi and Leena were approaching the base of the hill and with a slap of the back of her hand on Xavi's chest, Leena directed him to follow her towards the fields instead of the forests into which Xavi wanted to continue his Mermai journey.

"You don't want to go in there," Leena said to him.

"But it's a simulation!" Xavi said in surprise.

"Trust me, not the forests. I did that on the last simulation."

As the two sphere walkers passed the dense forest area, Xavi was provided his warning:

> The forests of Mermai are notorious for their intense gaseous nature, with trees emitting a strong natural chloroform, which, when combined with the heat of Mermai does not degrade; rather, its strength pacifies most entrants to Mermain forestry areas.

As Xavi and Leena deviated towards the fields, the narrator altered tack to reflect the changed environment:

> The Merman mint herb is similar to Earth in its genus but variance of water and the super-planet characteristic ensured Mermain mint is larger and more intense. This is a common aspect to all Mermain food products which tend to be greater in potency than their like-genus types on Earth.

Olisa took a white padded seat next to Hames. "She's quite bright isn't she?"

"Leena?"

Olisa looked back at Hames unflinchingly. "Not Leena. I think you know who I am talking about."

"Yes, okay. Emmalyne." Hames was right, an unavoidable conversation was about to begin. "All her marks from OUF Preparatory College say she is very intelligent and so far she's been fine with all her calculations. But she's young, probably too young. I haven't asked her what made her take the decision to join the Medens, not my business. I don't know, most kids her age don't choose ten years in deep space as the way to spend their formative years."

"Yes, as I said to her before, she is quite fascinating." The older crewmate shifted in her seat and spoke while still looking at the sphere holding Leena and Xavi on their trip through Mermai. "Emmalyne seems an inquisitive creature."

In the pause of discussion Hames and Olisa could see Lena and Xavi had begun walking alongside a stretch of beach that ran as far as the eye could see:

> Mermain beaches and rivers were protected from settlement at the outset with the first kilometre around them declared "parks of interest". As such settlements were built further inland. With hover-carts and space docks the spread of humanity that had occurred on Earth did not require replication.

"So how is she doing, Chief?" asked Olisa

"She's fine, just fine."

Nodding slowly, Olisa made the statement Hames was expecting. "Chen Chung told me about her moment in the repair engine room, just before we entered light alignment mode. It sounds like she was quite distressed."

Hames grunted while he looked through some calculations on a tablet screen. "While I know talking about things is the best approach for stress, I'd suggest it's better Emmalyne talks about her problems, not Chung."

"Chen Chung was only showing concern for her, he is an engineer after all. He sees a problem and he sees a solution, a way to fix it."

Hames could see that Olisa meant herself as the solution. An assertive tone started to creep into his voice. "I'm talking to her, listening to her, that's all she needs right now. She's not an engine part."

"Oh, but humans work just like engines. Brains are computers after all and they run our engine. We just need to work into the mind to see what is wrong, a little shift here and there and we can all feel peace. We are everything after all."

Hames did not say anything. He really wasn't prepared for where this conversation was heading and certainly not prepared for Olisa to start quoting Universal Direction at him late in the evening.

The old woman shifted in her seat as if she knew something clear and certain that had to be shared with her colleague. "You know, I think she may be of age, Chief, the age at which she can truly touch everything."

Even though he knew it was coming, Olisa's comment hit Hames like a ship at light speed smashing into a moon. Emmalyne, like most young people on Earth, was prevented from experiencing the second stage of the Universal Direction approach to understanding and appreciating the world around them; mind healing or Curate-mente as it was known. This prohibition on accessing young minds was a core tenet of the Uni-D movement – that children's heads were not to be touched by mind healers. It was considered that the brain took time to form and that meditation and openness of the heart was enough for the young. Hames regarded this as a positive point. However, the fact that a mind had to be settled to accept Curate-mente was a cause for concern.

"Listen, Emmalyne has just come out of stasis. So she is up and down while she sorts herself out." Hames shifted uncomfortably in his seat. "During the time in stasis all sorts of chemicals are pumped through you to regulate the body, and she's still detoxing from these. I know you think this is the perfect circumstance which calls for your sort of work." He shook his head. "I'm just not sure Emmalyne is capable of making such a decision right now. As her superior officer, I am asking you to leave it alone for the time being."

"Chief, I know you don't agree with the benefits of Curate-mente. You've evaded me for so long on letting me talk to you about it. Can you at least tell me why?"

It was a fair question from Olisa. Why was he so afraid of letting someone else touch his mind? How did he know that he wasn't broken in some way and that he might benefit from it? Perhaps he wasn't in need of assistance, perhaps nothing would happen if he opened his mind up to Olisa. Maybe mind healing only worked on people who really needed it. He never tried it and this was always a stumbling point in his arguments with healers, although he was far too stubborn to ever openly admit this.

In the silence that followed Olisa's question, Leena and Xavi finally reached the Bullet River.

"I'm a bit nervous. What's it like?" Xavi apprehensively asked his colleague.

"Don't worry, you will love it. You'll feel a little frightened at first, then it will feel normal and exciting to be in the fast lane!"

The Bullet River was named for its multi-dimensional currents that give the effect that in certain eddies there is no water, rather a stream of bullets being shot through the water. Eventually this phenomenon was better understood and now each year on the nadir of the annual summer festival, representatives from all over the planet with various types of crafts attend to compete in the Mermai Bullet River Championships.

Hames could see the sphere warp from the outside to create a feeling below Leena and Xavi that must have felt like kayaks entering the river. A moment later he heard Xavi scream and the screen revealed the terrified face of the repair engine officer being hurtled down the Bullet River as Leena laughed beside him.

Olisa continued. "Chief, you are the smartest person I have ever met, smarter than Yolanda. I could not begin to understand the calculations and concepts that you are capable of. Your work on particle repair technology is unparalleled, it's as if yours is the only will to mend the Universe itself."

"Now you're laying it on too thick, Olisa."

"Yes, but I want you to understand that I could never touch that part of you, not that I would want to. Even if I could touch it I could not affect it, change it or alter it, let alone understand it."

"I've had that said to me before. I still haven't changed my mind."

"Well, hear it from me. Again, all I am here to do is to show you the way to harmony through giving your mind an opening. An opening to seeing that you are not alone in this Universe. You are with us always, everything and everyone."

As Hames and Olisa sat in cold silence, Xavi and Leena finished with the Bullet River and pulled up at the wharf of a large Mermai settlement.

The settlement group at the main migration point on Mermai originated from the south-western points of Asia, predominantly from India. At the last recorded point, Mermai's population was over four million people. Initial colonisation was so dominated by Indians that in 2338 a campaign was raised to change the name of the new settled planet to New India. However, this campaign failed as some said it was not in keeping with the spirit of new beginning associated with all new colonies. A small number of protests arose and these grew into riots...

"Olisa, I made my decision a long time ago not to take the step you offer. It's just not for me. I'm telling you this one more time." Hames turned around in a show of defiance, and looked Olisa straight in the eye. "Don't ask me again."

"I will not, as is your request. One day, Chief, you will come to me seeking answers and you know how I would give them to you." Olisa was done. She had marked her claim on Emmalyne and left Hames to consider it. She turned and bid him farewell with her normal refrain, "With light, Chief," and left him to his work.

After the "free Mermai" movement began and the signing of the Eagleton Accord, Mermain independence was achieved and Mermain culture began to flourish and grow even more with further migration from the southern Indian areas of Chennai and Hyderabad.

Feeling bruised from the conversation with the ship's healer Hames decided not to wait for Leena and Xavi to finish with the Williams Sphere and left the Rec Room to return to his home-pod. Curate-mente made his mind race with dissent. He had seen other young colleagues go through this. After the first time a mind was accessed for healing there was the potential for that person to change. Some of these young people changed only slightly, some irrecoverably so.

It was said that understanding where your consciousness sat in the Universe gave that person a greater maturity. Some people, after undertaking the new experience, could come away from it markedly different. Hames had observed a flat-lining of personality in a small number of people. An inhibition of conciliation and conservatism of serendipity came over them, which to Hames felt like someone had sucked the life out of them. While the process of Curate-mente focused a person on 'the big picture' of the Universe it also induced a disregard for the smaller things in life. Little things gave character and colour to people, and Hames didn't want to spend time with a person without colour.

Emmalyne had colour, lots of it, and he was building a resolve within himself to ensure she stayed that way. This wouldn't be an easy task. Most young people through convention and curiosity sought out a healer to take them into their consciousness as soon as they came of age.

Olisa wanted to take Emmalyne down this path, completing the rite of passage and access to the Universe through a liberty of consciousness. Hames was going to find out whether Emmalyne was interested in Olisa's offer; he wasn't now and never would be.

Chapter 9

Watching Light

FOR HAMES, ONE OF THE MOST joyous parts of working on the *Medens* was that it travelled by use of light alignment drives or LADs. This form of interstellar travel made him feel warm and excited, like a young boy again. Perhaps it was a romantic notion but he loved the old-fashioned nature of travel by light. As a technology that was over 100 years old it should have been consigned to the annals of history; for most of his younger years not one person had travelled by light alignment drive. The period of its use had even been referred to as the 'Light Alignment Age', a period of time bookended by its creation and cessation once particle drives were invented. Its reinvigoration was only required post the events of the Cataclysm.

Possibly, he should resent the presence of LAD tech. It was indeed a symbol of broken space, a symbol that man had grown too much, too quickly. In the Cataclysm aftermath he remembered much bemoaning of the use of light alignment drives; people had felt it was a step backwards. Hames, though, couldn't bring himself to that point of disdain. There was something enriching and beautiful about light alignment drives, in the technology and the visual aspect of their operation.

Waiting for Hames and Emmalyne outside the entry to the LAD module was the gangly yet strong Chief Engineer Yolanda Hollings. Hames had an affinity with Yolanda; he had known her since they travelled together to New Galapagos not long after the Cataclysm. Like Hames, her looks belied her true age. Save for a wrinkle here and there, her skin was perfect and while her hair was deep white-grey, its short crop accentuated her high cheeks and rosy complexion. She was the same height as Emmalyne and her poise was typical of a person who had embraced Uni-D and was happy within herself. Hames always observed her as someone who could be encouraging of those who worked for her yet still be firm – not rough or aggressive – more maternal than military.

With a faint northern American drawl, the chief engineer greeted them.

"Welcome, Emmalyne. I think you'll like what we do here."

"I think I will." Emmalyne produced a sly grin as she responded with bright hazel eyes.

Yolanda led them through the entry point that expanded its shape to allow them access to the LAD module. "Okay, obviously we're pretty important in here. Without us the ship doesn't move and you lot would repair less than a scratch of a scratch on the side of the particle debris." Yolanda raised both her hands in mock apology. "I won't go into the whole Mission versus Ops argument."

"Thank you," Hames said calmly.

"You're welcome," Yolanda nodded back.

As they entered the Light Alignment Engine module Hames admired its complexity. Shaped like an octagon, each side held a doorway to a function or exit point, such as the one they had entered through. In its centre sat a set of control desks, also shaped like an octagon with what appeared to be correlating controls for a number of tasks.

Yolanda motioned to the first door on their left as they walked past. "That's the battery room, yeah? Can you hear it pulsing with power? Some people say they can actually feel it. I can't anymore, I think I've blocked it out. Over here is our control desk and next to it there is a place for us to study schematics of the light alignment drive." They walked around the octagonal control panel. Emmalyne looking suitably impressed as she acknowledged the various complex reasons for each part of the control area. Once finished with her descriptions, Yolanda moved them over to a nearby plain desk area and introduced them to the sole officer at work. "Right, this is Junior Engineer Robero Paldini. He also came on board when you arrived, Emmalyne".

The blond-haired mission officer moved over to the young man to say hello; however, she received no recognition from him. The scruffy, brown-haired head of Robero Paldini was bent over massive sheets of laid-out screens showing pictures of such magnification that they represented parts of less than a centimetre in size. Robero was moving over and around the thin screen images in an almost rhythmic motion, making notes at various points. Before any pleasantries could be exchanged, he looked up from his work space. There was a pallid sheen to his lightly bronzed skin, which was complemented by misty grey eyes that were ensconced in ever deeper blackening rings. In his brief encounters to date with Robero, Hames had not realised how poor a state the young man was in. Emmalyne herself hesitated and drew in a breath when he looked directly at her. Robero moved his eyes ever so slightly as if trying to look into her, then he dropped his head in a jerking fashion returning to whatever work he was previously focusing on.

"What's wrong with him?" Emmalyne asked Yolanda.

The chief engineer threw up her hands inconclusively. "He's the same as you I guess; you know, stasis-affected. Sometimes he's a bit zombie-like, you know? A Paldini Zombie or P-zombie if you like." Yolanda then paused, looking back and forth between them. "Wait a minute, he's not the same as you, is he? You seem to be doing fine. Robero, not so much. He just can't seem to shake the tiredness. It's in his bones for now. The healer has been working with him, although I can't see it. Olisa says he's improved no end and that he is fine to work. More so, that work is the best place for him. He's a good lad and isn't splitting an atom about it, but I'm pretty sure his body is telling him to hit the sack."

Robero was not too far away from this conversation. The rhythmic movements as he hovered over his work deviated for a moment in recognition of this commentary then restored themselves to their original patterns.

Emmalyne sighed. "I feel like that on the inside; it may not look like it, but I do."

"Well, you do a great job at hiding it, darling, great job." Yolanda patted Emmalyne twice on the back then walked back over to the octagon-shaped control panel in the middle of the room. She appeared to be enjoying the induction process. "Needless to say, the light alignment drive is integral to your and our mission. It's very old technology, been around a while as you know, so it's very safe and proven. The only issue most people have with it is it's the second choice; it doesn't compare with particle drive tech when you're talking about distances. Particle tech can take us across the Milky Way in almost an instant, but the LAD just crawls across the galaxy. So we travel fast for humans but not fast enough;

nowhere near it when you're talking about the colonies." Yolanda finished with a tinge of regret in her voice.

"The chief and I have been talking about this a lot lately. The *Medens* is tasked with repairing one of the smallest tentacles of the PCD fracture. It's going to take a long time to do this at its current speed."

"You're right," agreed Yolanda. "It's low priority stuff for the Organisation. They're all Earth-focused now. They'll send us out here because we don't cost too much for them, and they'll put some credits into researching alternative forms of travel. Again, not a lot." The chief engineer continued a walk around the control panel, programming some sort of report back to a central screen that showed a code building. "Most people have just readjusted to limited interstellar travel. No-one really seems to be pushing for it to happen anyway; small groups, survivors from the colonies who lost friends and family. But it's been nearly forty years since the Cataclysm so the voices driving any more effort to research another form of travel like PCD are getting quieter and quieter." Yolanda looked at Hames, who was doing his best to not speak up and let the chief engineer explain her area.

"Do you know when PCD travel was last attempted?" Emmalyne enquired.

"Not successfully since the Cataclysm. I can't think of anyone since who has attempted PCD travel to locations you need to have mapped, like Mermai. The space simply can't support it. A ship would just disintegrate and there isn't much point in PCD travel into blank space. Back after the Cataclysm, some tried it and they perished instantly. One group made a PCD trip to nowhere and returned. They were eventually found out." The code the chief engineer was building was finally ready. "It wasn't a fancy parade welcoming these people back or celebrating their efforts. No; they were punished and punished quite severely."

Tension filled Emmalyne's voice despite her attempts to sound relaxed as she patted a hand on the desk next to her. "I read about this in the analysis. Did they really put them into long-term cryo? Really?"

"They did," confirmed Hames, who was feeling nervous about this line of questioning from Emmalyne. She had just been through her own period in stasis. Hearing about the cold-hearted pragmatism in the use of cryo as a penalty for attempting PCD travel had the potential to strike a chord too deep for her mindset of the last few days. He attempted to close out discussion. "I'm not sure we need to go back over this; it doesn't really relate to the LAD and how it works."

"It's okay, Chief, I'm happy to hear it. I did ask the question," Emmalyne reassured him.

"Well," Yolanda looked at Hames for approval and with his nod she continued. "Cryo was brought in due to some twisted logic that time locked in a cell was considered a wasted life. It didn't quite line up though. I read the analyses of one of the first long-term cryo revivals. As you might know, five years seems to be the safe amount you can stay in stasis. You, girl, were only under for two." Emmalyne winced at this memory while Yolanda continued. "The first subjects served sentences of ten years in storage and were then revived. Out of the first hundred done only one survived past brain death to live for a few days. That time was cut short, though, as the subject took their own life once they returned to Earth from the cryo facility on the Earth's moon. Thankfully the Organisation came to its senses and realised it was more barbaric than prison and banned it."

"Agreed," said Hames, looking at Emmalyne for a reaction, but she remained quiet and hard to assess. "Those first ten years after the Cataclysm were tough. When they got those people out of cryo and they all died the uproar nearly split Earth in two. It was horrible; like we'd all made a collective mistake."

The American tones of the chief engineer became more upbeat. "I reckon it was only because of Uni-D that humanity kept it together." Hames preferred not to argue this point with Yolanda, letting the light alignment history lesson continue. "I'm not sure how they would punish someone now for PCD travel. I just know they wouldn't use cryo. Enough of the depressing stuff; the real treat is ready, and over here in the observation room."

Yolanda drew them towards a floor-to-ceiling square panel of glass that showed beyond it a square and spartan room. The sole object within it looked like an engine of some kind, bursting from the middle of the floor. "See there, that's the top turret of the light alignment drive. We can enter the room from the door to the left, via a dark staging and equipment area. From this turret we can watch the real engine on the *Medens* at work." Yolanda winked at Hames. "The top turret is the closest we get to the LAD on a day-to-day basis. It's where we can monitor the key aspect of the drive; you know, how stable the alignment is."

"Oh my, I've never seen a LAD in operation. Just read books. How does it all work?" Emmalyne seemed more relaxed as they returned to the task at hand.

"It's a combination of things. There's the beam oscillator, which sits in the middle of the drive. It basically spreads a field of energy or light across the whole of the *Medens*. It starts in the drive engine below, but it's contained and made impervious through additional regulator points across the ship, so the field is strong and protective. It makes us all resonate at the same frequency of energy and light. It binds us all together. Well, it binds the outer casing of the ship together and to an extent all of the crew. It has to in order to take us along for the ride."

"You make it sound so beautiful, so harmonious!" exclaimed Emmalyne.

"Just wait, it gets better. So we're all at one with each other, in terms of a frequency of energy. We're an object of energy and light that is able to be shifted in order to align with different frequencies that energy and photons travel on. While we struggle to move or access the energy around us, the spectrum of light is another thing. It's vast and wonderfully available, from absolute darkness to the brightest light, to all sorts of frequencies we can't even see!"

Hames was nearly as excited about the explanation as Yolanda was. "You see, every day we look at brilliant stars shining towards us. They give us their image from thousands of years ago, billions of years ago. The same happens with the light we produce: it moves away from us, it just doesn't travel very far, it dissipates. But that doesn't mean we can't use what is ahead of us," he finished with a mysterious tone in his voice.

"What do you mean?" Emmalyne looked confused.

Yolanda continued. "Just as you shine a torch out, say across a field or a corridor, after a distance that light loses its strength, fades, spreads, you know what I mean?"

"I'm sure she knows what you mean," confirmed Hames with a nod, a slow smile coming from his blond-grey beard. She's pretty switched on this one."

"Right at the start of that torch, the light moving out from it, even if for just a few metres of so is moving at..." Yolanda stopped and raised her hands as if asking a question.

Slight hints of frustration started to creep into Emmalyne at the raising of the simple question. Puzzled she finished the sentence: "The speed of light?"

"So what we do is create a source of light, one that is strong and consistent, and we map it. We know what velocity it's moving at and we place another contact point at the end of this light source, to anchor it. So we have a solid consistent point of tracking on the speed of light. Then, we take the field of energy we've created around the *Medens* and tie its frequency into the mapped point of light, just a fraction behind it and boom! Off it goes. The *Medens* is basically like a donkey chasing the carrot, always trying to catch the frequency above it."

"Amazing," Emmalyne said with a big smile returning to her face.

"That's the simple version. The very very simple version. Enough to give you a picture. It takes some good navigation to make sure we don't hit anything. I guess that's what the long range scanners are for, although the energy field is pretty powerful so if we hit a small object like rogue rubbish from a ship we'd blast through it. A planet or an asteroid, no; that'd take us out."

Emmalyne still had her mouth open. "I'd love to know how they came up with this. Who was it who invented it?"

"Mal Kara, an Australian," Hames answered. "Over two hundred years ago. It took some time until it could be used for interstellar travel. After that, it was used until particle conversion drives in 2305. The Universe is so massive that even light speed doesn't get you that far."

"No, it doesn't, but it sure is pretty." Yolanda was clearly proud of her work and Hames noted she must have given this talk a few times.

"Of course, you need to run a full eye over it once in a while, yes?" Hames did his best to present this as a question although it was more a statement to show Emmalyne he knew something about what was going on in the LAD area.

"Yep, that's correct." Yolanda smiled, knowing he was throwing her a lead. "We don't get to do full overhauls that often. Those take a whole month and the most we can do are minor jobs in the days we coast when taking on new personnel or supplies. If we start to run for long periods of time without stopping, which it looks like we'll do soon, we'd have to plan to make a big stop every six months to do the big outage checks.

Okay, that's the introduction done. Are you both ready to have a good look at the alignment engine? My crew has just finished the outage work from the last few days so it's all in top condition. We just want to view the light spectrum coming off the turret. It's a small test and a spectacular one." She winked at them slowly. "More for show really."

The chief engineer started to guide them towards the turret room. "When we first go in and open up the portal it's going to be a little bright; put these on." Yolanda offered them dark lens goggles and they entered the shadowy staging area through a door to the right of the control desk. They all put on belts with circle devices on them and as Yolanda fitted them she gave an overview of the technology. "These basically cancel out any possible radiation, of which there should be none at that point given the design of the LAD. A precaution if you will."

The three crew members entered the room to what seemed to Emmalyne a fairly clouded vision. Her instinct was to remove the goggles as they simply didn't seem required.

"Keep the goggles on," Yolanda cautioned. "When we open the viewing portal it will get bright enough, trust me."

The turret looked like a big mechanical engine from many years ago. It held no distinctive shape due to various tubes of glass and metal pouring in and out of panels and metallic boxes. Near its zenith was a small viewing portal the size of a dinner plate, with a metal flap overlaying it. Hames was excited as he knew what was coming next: a beam of light that would shoot above the viewing portal to a set point on the ceiling. It would be like a small rainbow tower right in front of them.

Yolanda gave Emmalyne the detailed version. "The light that will come out of the portal is targeted at the sensor plate above it. It feeds back readings to the workstations, and the data from the sensor plate tells us various things about the performance of the drive. It should also show us a very beautiful and distinct multi-spectrum light stream."

"Is it safe?" Emmalyne looked concerned.

"Absolutely, it's fine!" Yolanda Hollings began to point at some of the tubes and squares. "As you can see, we have regulators and buffers around the base to ensure the sample of light redirected from the energy field around the *Medens* is harmless. It'll just make the room brighten up somewhat. Right, let's have a look." Yolanda lifted the flap to show the beam from the light alignment drive pulse out of the viewing portal and instantly illuminate the sensor plate above it.

Hames walked around the room, watching the colours in the vertical beam shift in his perception just as his position shifted. He looked across the turret at Emmalyne and she seemed at peace. More relaxed than he'd seen her face for days.

Suddenly, in the corner of his eye, Hames noticed an unwelcome shimmer move out of the tall beam of light and flicker across the room. He looked around the turret. Everything seemed as expected. Then he looked below. One of the buffer tubes appeared to have a scratch on it. Hames first considered this an inconsequential fleck, reflective of the last three years of operation. He thought not to mention this to Yolanda. When he looked again the scratch was now a crack and the technical experience inside him recognised that everything was about to move on from being fine.

"Emmalyne, close your eyes!" Hames shouted as he moved to the buffer at his feet. It was too late. The tube came loose and swung up to become another vertical shaft of light, pulling a linked connection point loose. Hames dived to the machine and pushed the tube back down. At the same time, a tube near Yolanda creaked and light burst from underneath it. Then an explosive force threw her into the back corner with a thud and brightness filled the room. Hames and Emmalyne, on opposite sides of the turrent, were also thrown violently back towards the walls. Instantly an intense light beamed throughout the turret observation room. The energy and frequency of the alignment drive was escaping and drilling into Hames and his crewmates. He was immobile, feeling warmth and light compress all around him. Briefly, he supposed this is what it must be like on the sun. Then he blacked out.

When he regained himself he wasn't sure how long he'd been unconscious for. He was totally lost, with his goggles askew, colours and brightness engulfing him. He felt as if he had lost his eyes and could neither close them nor open them. Hames started to acknowledge heat all around his body, although, his confused senses and a ringing in his ears made him feel immensely numb. Disorder reigned in his head; he experienced fractals invading his being. "Stay," his body sang to him. "Stay in this," he heard, and he did want to stay.

But what about Emmalyne, Nisha and the *Medens*? There was Emmalyne somewhere else in the room, invaded by light. He must move and find her. The brightness of the alignment drive continued to permeate aggressively around the room. In the present, in the now, in the space of warmth, he tried to think. He told himself to *get your bearings*, though this was easier thought than done. It was like trying to wake from a deep sleep on a lazy Sunday afternoon. His body was warm and he so desired to stay beholden to the light. He tried to open one of his eyes yet nothing happened; blindly he drifted back to the warmth. Hames pushed himself again. With immeasurable effort both eyes released. Once his lids were prised open he found the next steps all the more achievable. Surprisingly, he could look into the colour and the light without a reflex action to shield his eyes. In a moment, that on later reflection was lunacy, Hames squinted over to the viewing portal stream that was still beaming at the sensor plate above them. From his position flat against the wall, he saw in the stream something different, maybe his mirrored reflection. It was a face shaped like his yet not his. Then it was gone.

He had to move his body. He had to help his colleagues. Thinking of Nisha instantly grounded him. He remembered the compasses on his hands, broken and unbroken. He focused all his efforts, and they lifted off the wall. Obtaining a greater level of conscious control, he replaced his goggles and the feeling of warmth was reduced.

With his hands awoken and free of the light, the rest of his body came easily to him. Finally, Hames fell forward off the wall. He grabbed onto the turret and staggered around to the other side of the room, making the assumption that Emmalyne, who had been in the opposite position to him, would have also been thrown back to the wall. Gingerly he felt his way across and there, as predicted, was Emmalyne. He grabbed her by the waist and was ready to pull with all his might to tear her away, but she fell slowly over his left shoulder and onto his back. He carried her in his strong arms to the door into the staging area. A darkness fell over them and he dropped to one knee and eased Emmalyne back over his shoulder, catching her head as he rested her on the floor.

Yolanda was still in the turret observation room but there was no possibility Hames could go back in there. It felt like someone had hollowed out his insides. He pushed back to a standing position and walked over to the door of the octagonal LAD module. He stumbled back in to the main control area and it was still plain to see that something had gone horribly wrong. The glass wall to the turrent room intensely pulsed white and yellow while an alarm rang out.

Hames looked around. Where was Robero? Why hadn't he helped them and closed the turret? He staggered about as he accessed the first terminal on the control station. The alarm protocols were there and waiting, directing him in easy steps to close the turret off. The light dimmed and then it stopped. As it did, Olisa and Francois come running into the LAD module straight to Hames. Without words he brushed them aside and pointed towards the direction of Emmalyne and Yolanda. They acknowledged his gesture and headed towards the critical patients.

Hames turned and in the back of the room he could see Robero crouched over on his haunches holding one of the large thin screens with the drawings of engine parts. He had it wrapped all around his body and over his head, rocking back and forth.

As angry as he was with the young engineer there looked to be a method to the madness. He slumped to the ground, grabbed the nearest screen, drapped it on his face and closed his eyes.

Chapter 10

Battle of Caltrap

FOLLOWING THE ACCIDENT WITH THE light alignment drive, an uneasy feeling had persisted inside Hames. This was noticeable once again as he swam up to wakefulness in the late afternoon. He had sat through many meetings in the last two days in which the COO and the ship's manager, Siggy Conson, pushed Yolanda and himself for answers. Amongst tired thoughts his remaining guess was that the viewing turret had somehow cracked under stress caused by the alignment drive. Some of the crew thought it an unlucky coincidence that it happened when he, Emmalyne and Yolanda had been looking at the light spectrum surrounding the ship. To him it was a weak and unscientific explanation. In the absence of more information, it would have to suffice.

Dragging a white t-shirt over his tired shoulders his hair was momentarily scruffed before he readjusted it to its customary neatness. Despite the event of blinding force that had knocked him around the turret room, Hames was untroubled. What was most important to him was finding Nisha. To do this the ship must remain on mission and repairing space, and at top speed. Satisfyingly, the COO and Yolanda had not turned the light alignment drive off. The *Medens* still continued on its way, repairing space, pulsing through the Universe at near the speed of light. This was a sensible decision. No critical damage to the ship could be discerned except for some small damage to the viewing mechanism in the turret room. Some of the crew who had seen the physical damage done to Emmalyne and Yolanda were uncomfortable with the decision; both had to be stretchered to the medical-pod for pulled muscles and broken ribs to be attended to. Hames noted, somewhat guiltily, that this disquiet had been tempered by the practices of Universal Direction, its daily meditations and conventions.

As Hames attached his work overalls he noticed some small rips in the elbows. It must have happened the other day in the LAD turret room but he'd been too distracted to notice them. The only other pair was in the washing stream, so he tied the upper half of the blue overalls around his waist, revealing his white t-shirt clinging to his ageing body. Now he was more awake and cognisant of the world around him it was time to visit the ship's store for new clothes before going to see Emmalyne.

Hames arrived at the store to find Francois and Leena moving various boxes around the wide, tall room. They seemed to be looking for something. Large two-metre square crates were being rotated up, down and sideways by their bare hands and a form of magnets against the back wall. The desired crate appeared to be located at the furthest point in the room. They paused when they spotted Hames.

"Chief, how's it going?" asked Leena turning around and dusting her hands. "You feeling alright now?"

"I'm okay. Not as bad as the other day; the head has stopped spinning now." Hames rubbed his cheek for a moment.

The large Frenchman, Francois, turned from the shifting of crates. "Has there been any development on explaining the accident?"

"I don't think so," Hames confessed. "I can't really understand what caused it. We're a bit stumped. The viewing turret shouldn't have broken like it did."

Francois' deep tenor was sincere as he lay a large dark hand on Hames' shoulders. "I know we are going to keep moving on to New Galapagos. I think that's a good thing. The past is past, and what will be will be."

Hames winced for many reasons. At the heavy hand on his sore body and at the blatant mention of not one but two Universal Direction conventions. In his stomach where an answer should be for how to respond to Francois, there was nothing except his desire to keep looking for Nisha. Redirection was the key; change the subject.

"So, what are you two doing in here?"

"Leena here caught me in the corridor near the bridge this morning and asked about my personal history. Whether I had fought in the Battle of Caltrap."

"It's true. I'm fascinated with the history of the colonies. I'm Mermain through and through even though I've never been there. I must be compensating." Leena sighed.

Francois shifted another crate in front of him to find his target. "I think I can find something that will give you an answer to your question." The grey hairs on his dark hands stood on end as the powered crates spread at his command. "Here it is: my personal archive!"

"Great! Now can you tell me what you are looking for?" enquired Leena.

"In good time. First let us revisit the period of my youth, when I was a young man living in one of the northern continents of Cernunnos! Pull up a crate if you dare, Chief Hames!"

Hames decided to hear Francois out. It was a pleasant distraction from all the chaos of the light alignment drive accident and Emmalyne wasn't going anywhere while stuck in the medical-pod. He joined Leena in leaning against a crate opposite from Francois, relaxed and alert; a willing audience member.

Francois began in his unmistakable tone, a voice still rich in its French- Cernunnosian heritage. "They were two magical members of the Phoenix constellation, dancing in tandem around each other. Cernunnos as you know was very ordered and structured. How could it not be with its chessboard-like layout of continents and islands? Lugus was very different. Rugged and hot with a little too much volcanic activity. A young and risky planet."

"With that risk came reward right?" started Leena. "I studied Lugus at college. It provided precious metals to all the colonised planets. More than all the mines on all the other worlds put together. The pictures of the open cuts were amazing. Like giant impact craters, but no meteor had ever been there."

Francois continued. "For decades Lugus was a colony of Earth. The Lugurians, though, did not think they got much in return for the resources provided by their planet. But this was the deal. They chose to live there and knew what they were getting into."

Hames could not help but join in. "It was the way with all of the colonies. Some people went there for reward, some for a new life. Some just to escape their life." He paused to scratch his arm under the white t-shirt. "For all of them, though, once a second generation came through and was born there, there came a sense of belonging and pride. All colonies eventually want to show their own identity and protect themselves, especially if they believe they're on the wrong side of a trade."

"Ah, Chief! Be careful there!" exclaimed Francois. "The Lugurians *thought* they were getting a raw deal, but it wasn't necessarily so."

"Weren't they?" challenged Leena. "Their planet was being dug up and all they were being sent in return was the stuff nobody really wanted. Old tech, used ships, just small quantities of the parts required to operate life systems. Plus, the Earth supervisors managed the delivery, making sure the best parts went straight to mining operations first."

Francois opened his face and hands in explanation. "Those metals were essential to the rest of the colonial system. Travel and trade to the other worlds would have collapsed without the resources of Lugus."

Leena moved forward. "And instead of giving them the recognition they needed, they were given nothing. The coal miners of the 23rd century, one analysis called them."

"Miss Leena, I am very disappointed that you can only see one side to this story." Francois' words suggested an unhappiness with Leena's position, but his tone and general calmness were testament to his Universal Direction commitments. "You see, it was not Earth that struck first, it was the Lugurians who took the Earth supervisors hostage."

"They cracked! Can you blame them?" Leena's voice was raised and her face scrunched. "They just wanted tools to farm, make more food processors, to heal their families. Then the Lugurian flu outbreak happened; such a hard life!"

"And that is why Earth took the Lugurian's particle drive mapping and ships away – to protect us all!" The unflappable Francois was beginning to show signs of irritation.

"Rubbish!" shot back Leena, banging down an arm on a crate. "They wanted to punish Lugus. If Earth couldn't have her resources, Lugus was going to suffer!"

"You are very anti-Earth. You were born there, no?" The burly Cernunnosian native appeared to be taking a different tack in the discussion.

"I'm not blind to history. Earth treated all the colonies poorly from time to time; it wasn't a good parent. For example, Earth abandoned Cernunnos when Lugus attacked."

"That was a precaution. Lugus wanted our PCD maps and engines. If they had got it the whole Universe could have been infected with their flu." Francois responded finding ground on which he appeared to feel more confident.

In the midst of this debate, Hames surmised that a different context could calm the conversation that was riling up the young first pilot. "Leena, people were scared. No-one over fifty years old survived the Lugurian flu. Young men like Francois joined the call to fight the Lugurians as they were fearful their friends or family may die, or they may have died, if the Lugurians landed on Cernunnos."

"Okay, I can respect that," said Leena as she leant back against the crate. "It just hits me that if Earth had thought more about the Lugurians at the start maybe they wouldn't have got sick. The whole situation could have been avoided. Strategically, Earth was short-sighted and hurt people in the process."

Hames hadn't broken down the Lugurian issues that way before; Leena had a view that was hard to argue with.

Leena came back to their original purpose for being in the store section. "So, Francois. You said you had something to show me?"

"You mentioned what you thought was the most poignant battle you had read about, the Battle of Caltrap."

"Yes, the first battle in ships and exo-suits across an asteroid belt no less. It sounded heartbreaking what the Lugurians had to deal with as well as how the Cernunnosians responded. The footage is just spectacular – I've watched it a thousand times."

In response, Francois pulled out a matt-black one-piece protective suit.

"You were a Caltrap stealth trooper?"

"That is correct." Francois' face was more ashen than excited. "I was one of the sixty-four men and women who broke into the night and planted magnet mines on as many asteroids as possible."

"Oh geez," said Hames.

"Yes. Our intelligence had told us the Lugurians were going to creep through the asteroid field slowly as they believed we would never expect this. But one of our spies alerted us and we spent the night before seeding the field. The already spiked nature of the Caltrap asteroids was made lethal on a grand scale with all our mines, which drew the asteroids into their ships."

"Oh my light!" gasped Leena. "You were part of that?"

"Yes, indeed. The Lugurians committed most of their forces to the attack through the field. When they were close enough the magnetised asteroids flew into their ships. We destroyed over eighty per cent of their forces and fought off the rest in a one-sided battle through the debris of rocks and their fallen friends."

Leena now looked concerned, in contrast to her former curiosity. "Weren't you scared, Francois?"

The burly French-Cernunnosian breathed out. "Of course I was. There was no time to return to our ships without alerting the Lugurians so we had to stay on the outer parts of the field, floating in the dark. We all thought we would die, from explosions, rock collisions or just from running out of air. I lost many friends. Back then I assumed I would die and pass into nothingness. Now I know better. We are all connected."

"I'm sorry you had to go through that. It must have been horrible. Before, what I was saying about Lugus," Leena struggled for words, "I just think it's a terrible situation Earth put both of the worlds in." Leena gave Francois a hug, her arms nowhere near reaching the full way around his belly.

Francois returned the embrace. "I left for Earth not long after the battle. I wasn't proud of the damage I caused. I was proud, though, to defend my world. It was meant to be so I do not regret it."

Hames uncrossed his arms and moved from the crate, leaving Leena and Francois to finish their discussion. He grabbed a new pair of overalls from the far side of the room and departed to visit Emmalyne in the medical-pod.

As he approached Hames could hear laughing so he stopped short of the open door. Not wanting to intrude he started to walk away. However, indecision took over him and he paused to rest against a wall. Never before had he been this unsettled. The LAD accident was affecting him and more than just the explosive photon slap that had thrown him across the room. At one point during the experience he was unable to move and had felt completely content being constrained. He had felt outside of himself, or was it just a reflection of himself? It was hard to tell.

Emmalyne's visitors, Temett and Dao, exited the medical-pod to see Hames. "Are you okay, Chief?" asked Temett with an inquisitive grin.

"I'm fine, just waiting for you to finish with Emmalyne."

Temett seemed to accept that no true explanation was forthcoming. "Sure, we're done now, you can go in."

"Thanks." Hames pushed off the wall and adjusted himself.

Emmalyne immediately spotted him as he entered.

"Chief! I'm so happy to see you!" Her blonde hair, loose of its ties, swung as she turned to him, and her rich hazel eyes sparkled. If not for the colourful bruise on her right cheek, Hames thought she looked fine.

He pulled a seat close to her bed and sat down. "How're you going with all of this?"

"Oh, you know, like a space worm ate me and squeezed me in all directions. I was pretty torn up in spots but the med-bar has tackled the bigger bumps. I'm still sore all over. It was like a giant hand swatting me away."

"Yep, It was a big push backwards. We're lucky we didn't break anything."

"Not Yolanda though. Temett just told me her arm broke in two places and she's got a massive concussion."

"She took the worst of it. She's tough you know, already out and about working."

Emmalyne looked like she was trying not to smile. "Out of the corner of my eye, while I could still open it, I saw her bounce off the corner of the room like a toy doll. It was pretty distressing yet weirdly interesting." Emmalyne put her hand on her mouth to stop the smile presenting itself. "So unreal, the whole experience."

Hames also did his best to contain a smile. "It certainly was different. How's your head?"

"It doesn't throb, not like the rest of me! I just have these moments where I feel a bit distant. Olisa tells me it's shock."

It was only natural, and required, that Olisa see Emmalyne to provide assistance to her medical situation. He didn't like it all the same. Since Emmalyne had arrived Hames felt a need to look after her. He had assessed her immediate dangers and threats. These had firstly been Emmalyne herself as she adjusted to life after her long period in stasis.

Hames recalled his conversation with Olisa in the recreation room and the suggestion that Emmalyne undertake mind-healing. It wasn't his place to get involved in this decision. However, history had taught Hames he wasn't good at letting go of things he cared about. "Tell me, what did Olisa have to say?"

"She said that I'd undergone a traumatic experience, that all three of us had. She said the intensity of the emissions from the LAD turret could have affected us in ways we mightn't know. She's offered to 'have a look' in her words." Emmalyne said this last part very matter of factly yet Hames knew exactly what it meant. Again he remembered the change in some of those friends and colleagues who had undergone Curate-mente.

"Can I ask you a personal question?"

"Chief, you've got my file. You know everything about me, even what I am supposed to like and not like. How I react to others, all that stuff is in there I assume."

"It is, and your particular case is fascinating reading," he said while grabbing his beard. "I know you haven't had Curate-mente before...well, what do you think about it?"

"It's something I've considered, sure."

"Have you ever considered whether it is something you should do?"

"Why would I not? I want to feel more in touch with myself and the Universe around me. People seem so calm and relaxed. To me it's like they have been given the answer to any question or any concern that could be asked of them. I want that."

Hames leant forward. "I never got formal tuition in Universal Direction. You've got to remember that I was twenty-one when Uni-D really took off. The meditation schools came later, as did the mind-healing. By this stage I had set ideas about the Universe. Hey, I'd mapped the damn thing! That was my job!"

"I get it. You were too old, too set in your ways." She grinned at him. "It must be hard being a refugee in a new generation."

"You're making fun of this old man."

"Just awful. Like Cro-Magnon man or the apes, you're a throw-back to our ancient forefathers." She smirked at him warmly.

He could see the funny side of her teasing. "I probably should be in a museum. I certainly don't think like most of the people on this ship. There are two sides to it, you know."

She looked back at him seriously. "What do you mean?"

"For some people there is an element of mind-healing that moderates or subjugates emotion. I've seen Universal Direction start from something small to become something that dominates our society. I've watched it develop, seen its up and downs. The times when, as a practice, it wasn't as refined as it is now."

"That's normal for anything, even an idea. I read about times when people used it for personal gain when they didn't really believe in it."

"I'm talking about something else. I'm talking about Uni-D changing people, and not for the better, for the worse. I've seen friends and colleagues change, to be really altered. Emmalyne, I don't want you to be lost too."

"I won't be, Chief, you know that. But it's my decision."

It was her choice, no doubt. In his view, though, she wasn't ready to make this choice. "Listen, it's going to be a good month as we approach New Galapagos, why don't you sit on it till then? In that time you can make a decision about what you want to do. You're not long out of stasis and after the accident with the LAD viewing portal, well, I don't reckon it's a great time to mess with your head."

"I take your point, Chief."

Emmalyne sat up some more, reaching to the table for some water. She took a drink and looked down at Hames' hands. "Where and when did you get those pictures?"

Hames turned both hands over and inspected them like he was reading an old book from his younger days. "I got one on Earth, the complete one. I was about twenty. Young and wide-eyed at the prospect of seeing the Universe. When you're that age and you look up at the stars, and someone says, 'you can go there' it's an amazing starting point. That's the context of my education. I began my life when we'd been using PCD tech for decades. I was barely sixteen when I got my first scholarship; they sent me to see all of the colonies, Lugus, Cernunnos...oh, but not Sol-Concio."

"Why not that one?"

"In truth, I don't know. I think the scholarship didn't include it as it's in the most precarious of the PCD routes. It's a tricky route, the system of planets is right in front of the Orion Nebula. So the travel to Sol-Concio was generally infrequent and, from what I understand, four years after colonisation, open migration was stopped. Not long after that, visits weren't encouraged, nor were departures from what I gathered. By the time I got around to wanting to go there the place wasn't really a tourism destination and not much information came out of there. It's nearly thirteen hundred light years away, it's just too far."

"Why wouldn't they let people leave?"

"Back in the day I heard some rumours. I'll spill them another time. For now, you need to relax, get some rest."

"Hey, not so quickly. You haven't told me about the second compass, the broken one."

Hames surprised himself by the feeling that came over him. It was a readiness to share and explain. "It's a reminder for me. A reminder that no matter how clever we think we are, how much we think we're running the place, anyone can get lost."

"I guess anyone can be lost, it's not a unique characteristic." Emmalyne smiled at him and for some reason, despite the weight of the hand that carried the broken compass, he felt lighter.

"You're right, you really are." He looked at her quizzically.

"What was it that broke the compass, Chief?"

"Not something, someone. We met on Mermai, on one of my scholarship trips. Nisha was a liaison officer at the space docks on Mermai. They were pretty friendly at the docks, in spite of the misgivings about interstellar trade and migration. I was a young engineer, in my late teens but I had already been through the Agency Academy once for my initial course. I was meant to stay there a week, gathering data on the galaxies past the Mermai system. I stayed two years.

"Chief, I would have never have picked you have a heart!"

"She was this amazing person who could talk and listen to me about my work and then take me completely away from it all on a quest for the simplest of things. We'd go from debating the importance of post self-determination constitutions to flying halfway around the planet to a chocolate shop she'd visited once as a child. Nisha taught me love and gave it to me, unconditionally. She even made me more passionate about my work than before. Eventually, the Agency asked me to return to complete my study on mapping new PCD routes. It was my passion and she let me indulge it." Hames did his best to manage the growing heat in his face. "She was so good to me. When she agreed to move to Earth for a year I felt so bad taking her away from her home. We were literally only days away from moving back to Mermai when it happened, the Cataclysm."

"I don't get it, what happened?"

"She was just gone, lost in the event." Hames was not feeling so open to talking about this part. His fist was clenched tight and his face showed the subject was not open to further discussion. "I'm sorry," he finally said. "It was a hard and frustrating time. It doesn't get easier to relive it. I know others had pain, worse than mine. It doesn't matter, though, that other people suffered. It doesn't help me in any way." For a moment, there was silence between them, allowing Hames to regain his composure. In the pause, someone approached down the corridor. Olisa came around and saw them, stopping at the entrance point to the med-pod. Hames shifted his seat back towards the door. "It's okay, Olisa, I was just about to leave."

"Please, take your time." Olisa turned from the entry point to let them finish their conversation.

"Emmalyne, do you know why they call the ship the *Medens*?"

"No, I've never asked," Emmalyne responded mildly, astonished at herself. "I'd always assumed that it was done in honour of a wealthy donor from Earth or an admiral, or an admiral's son lost in one of the outer colonies."

"Those are fair assumptions," replied Hames. "*Medens* means 'healing' and as we're essentially healing space it was said to be appropriate. I guess we have been healing ourselves ever since the Cataclysm."

"You don't believe in Uni-D, do you, Chief?" Hames was unsure what to say. He didn't think his answer was what she wanted to hear. He also sensed himself being watched by Olisa. Even though she was in the corridor he could feel her eyes on him, waiting for his response.

"In my background, things need to be proven. Shown to be positive, negative. You know, empirically tested. The idea that consciousness is in everything and is everywhere just doesn't ring true with me."

"Chief, you can't take a reductionist view just because we can't prove there is more to life outside of us. It's limiting to suggest there is no connection across worlds, across time."

"Like I've said before, show me the proof. Give me the test results and I'm there."

Emmalyne's voice began to rise. "Consciousness is atemporal. How could we measure it scientifically? The Cetana Initiative spent years before the Cataclysm affirming the non-local nature of our existence."

Hames stood up. "Right, you. You need to rest; get some sleep to heal up."

Emmalyne flopped back in her bed. "Sleep. Sleeping is the last thing I want after the last two years asleep."

"Tough talk from a lady who needs to stay in bed."

Without looking back at where Olisa was waiting, Hames turned left down the corridor. It had been a good discussion, although he still felt something nagging at him. It wasn't that he'd upset Emmalyne at the end of their discussion; that seemed normal enough. He had a feeling that the whole incident with the light alignment drive wasn't an accident. Maybe it was another thing altogether.

Chapter 11

Mess Hall

HAMES WAS ALONE IN THE repair control room. Time had sped up and moved past his methodical meandering. The days had been long as he covered Emmalyne's absence. Now it was early evening and past his customary time for nourishment.

As he approached the mess hall, he heard the familiar buzz of voices and noise that suggested a number of people were within the earthy-toned room. The evening's Universal Direction session had long since finished, meaning the mess hall was busy. The noise grew louder and louder, overcoming the beat of the energy conduits beneath him. At the entrance point he realised the whole ship was present, even Emmalyne and Yolanda.

Hames looked for a location where he could sit by himself. Typically, this meant a space at a smaller table on the far side of the room. Unfortunately, all the seats in this area were taken. In fact, all the half-quiet options were taken. With a sigh he made his way to a seat in what he considered the worst location, in the middle next to the utility wall. He put down his tablet with his calculations and accessed the dispenser.

Tonight he felt the need for salt, carbohydrates and a rush of nutrients to energise him. This called for an Cernunnosian salt cake, his go-to solution in times like these. Although not many exo-planetary foods remained on people's menus after the Cataclysm, the delicious and mouthwatering nature of salt from Cernunnos was irresistible and not too difficult for utility walls to reproduce. To complement the salt he needed sweetness, so he requested his usual beverage, a Mermain hot chocolate.

As he sat down, he realised that the other two members in his section would soon be joining him after collecting their dinner. Olisa and

Chen Chung looked deeply engaged in some discussion that was distracting them from accessing the food they required. Hames pleaded internally that they sit elsewhere. He even reached out with his mind, thinking there must be a shooting star, somewhere outside, that he could wish on. The outcome was unavoidable, however, as the four-seat table was all that was left to share.

Chen sat down first. "Hello, Chief, we were just discussing tonight's session. I had the most amazing meditation. I really felt part of everything around me. Everything I'm seeing...my sensory acuity is just incredible!"

Hames could not even muster a 'that's great' or any of the other non-committal answers he employed at these times. He simply stuffed as much Cernunnosian salt cake into his mouth as he could. Olisa then sat down and gave him a disapproving look before continuing her discussion with Chen Chung. The chief mission officer took this as an opportunity to socially retreat and pulled out his reading tablet to focus on research, and he began flicking through the data on the Mermai Tentacle, looking for ships that would have travelled this way nearly forty years ago. His concentration was tenuous as he was assaulted by the noise and smell of the environment around him.

He dug his head down and tried to focus. This evening he was revisiting the pathway through the Mermai Tentacle of the *Python*, the ship that started the Cataclysm. Years ago he had learned that Nisha had been on the *Python* that day. It was immensely complex work and proving impossible in the lively room. He was being sporadically bumped as crew members went back and forth from their tables to grab more forms of sustenance. His accidental data entries were a signal that he should focus on one task alone, and that was eating.

The COO rose from another middle table in the room to get their attention. She did not need to seek their quiet; the mere presence of her standing tall above was enough to silence them.

"Thank you for your attention." Melini Guerra's strong, calm voice held the room, her hands on hips as she breathed out. "Some of you know that today is a special day in the Mermain calendar, it's Diwali. It's taken from one of the older practices and given it's about light triumphing over evil it feels a lot like what we're doing here. Healing the black of space to bring light back to the colonies we lost." A few of the crew gave their leader noises of agreement and support at this comment. "Normally at Diwali we have a few celebrations." The COO pushed her hands out sideways in mock apology. "I don't have fireworks or even Ugor Eels to offer you as a spectacle but if you'd like to join me in the quadrangle later we can take some time to watch the stars flying fast above us. It's a pretty good light show in itself."

"Absolutely, cousin, I'll be there for sure," called Leena.

Melini Guerra brushed her hair backwards with her hands to clear her face. "To help celebrate Diwali, I've prepared some ladoo and halwa, traditional treats from my home. I'll bring them around for you to try. Thank you and enjoy the rest of your meal."

The crew of the *Medens* gave their leader a round of applause; even Hames found himself clapping and he was happy to do so. He could see that Olisa, across from him, was not so enthusiastic. Hames leant across to the old healer, raising his voice to get above the final claps. "You don't approve?"

"Nonsense, Chief. Universal Direction is accepting of all paths that lead to light. We are all everything after all."

"Great, just checking." He stopped clapping and looked around the room. In the far wall areas on the smaller tables were Temett and Dao Mai, the latter working on a tablet with greater success than he'd had. To the table next to him were Francois and Siggy Conson, planning out shift rosters for the coming days. In the very middle of the room at one table were Xavi, Yolanda, Emmalyne and Leena. The two females appeared to be facing off against each other in what seemed to be a game of oneupmanship. The most Hames could glean was it had something to do with who received the most reprimand notices or 'red flags' at OUF Preparatory College. Hames remembered red flags – the official records of discipline incurred during a person's time at college that were typically accompanied by some form of additional duty.

"I don't care if you had five and I only had three. All of mine were given to me by the Head of College, so that makes them more serious than yours!" Emmalyne protested to Leena with a large grin.

Leena's pointed nose scrunched in her bronze-hued face. "Argh, it's basic maths, five is greater than three, deal with it!"

"I never believed that you'd have such a simple point of view. I had to spend hours editing star charts in the Head's office? Summarising algorithms for launch trajectories!"

Leena sat up straight and relaxed her shoulders. "One of my five red flags was to do with a visit to the Head's office."

"Ha, you see! Most of your flags were all at the hands of minor tutors, small-minded twerps! Admit it!"

"But," Leena continued, "it wasn't the Head who gave me the flag."

"I'm confused; you said you visited the Head's office?" Emmalyne asked

"I visited the Head's office, not because of the flag. I went there while she was in assembly. I was to leave a report she'd asked for. When I got there I saw on her desk a stasis jar of vacuum bugs, you know the harmless ones. It was a kind of decoration, with lights shining through them."

Emmalyne looked a bit pale. "Yes, those are pretty aren't they?"

Leena went on. "I couldn't help myself. I opened the stasis jar, hacked the room settings and ramped the heat. A lot."

"Oh no, you didn't!"

Melini Guerra placed a plate of ladoo and halwa at their table while confirming her cousin's actions. "She did." The COO looked at them shaking her head.

Leena continued. "When the Head came back she couldn't even open the door. The vacuum bugs had expanded to fill the whole room and had jammed the door by pushing through the control panel to ruin the circuits. The Head had the room scanned and from the bio-readings they figured out what happened. By this time the materials in the room had all begun to fuse together; they had to use laser cutters to open the door." Leena pretended to look down guiltily although she could not pull off the act.

The COO finished the story for her young cousin. "I was called in before the Organisation's Board. The Board! I couldn't believe it. The Head of College wanted to press criminal charges, have her put in cryo. Fortunately, the Board put a fair amount of value on the *Medens* project back then so Leena wasn't kicked out, as long as she was kicked further into the Organisation as it were."

Leena looked at Emmalyne exasperated. "You don't think I'd choose to come out here on this trip to nowhere? I had to follow my protector here or get into massive trouble." A content look came over Leena's face and she closed her hands together on the table.

"The Board," said Emmalyne. "It's technically not a red flag. I'll concede, though. You win by some distance." Emmalyne flopped her head down in her arms on the table.

"Five red flags? That's nothing! I can beat that," chimed in Yolanda. In fact, I think I still hold the record: thirteen."

Emmalyne's jaw dropped in slow motion and her eye's widened. Leena looked at Yolanda as if she had just had first place in a cross-solar navigation race stolen from her. "That's not possible! You couldn't have completed college with that many flags!" She was clearly furious at her victory over Emmalyne being snatched.

Yolanda rubbed her arm, still healing from its break in the accident of just days ago. "Oh, I guess it was different in my day. I entered the college in its first years as they were preparing to re-enter space, about seven years after the Cataclysm. So I must have been, ah, about thirty-eight. I was the same age as the instructors so I questioned them on a lot of the rubbish they put out about drive engineering. When it got heated, the instructors, who eventually realised they were wrong, gave me red flags as a way for them to save face or massage their egos back into shape." Yolanda chuckled as she finished this.

"I was in the same intake with Yolanda at OUF and I can tell you some tales about my chief engineer." The COO turned to the two young bickering women and tied her hair back as she spoke. "The college was lucky she went through when she did. They lacked some serious knowledge about light alignment travel. If it wasn't for her a lot of the renewed generation of light alignment ships would have blown up the instant they started them up." Melini Guerra became particulary animated at this last description. "Despite the arguments and the red flags, in the end they all knew she was right; that's in spite of her cantankerous behaviour." The COO stopped and pointed at her friend. "And one more thing, she got seventeen flags, not thirteen!"

"Really?" Yolanda asked with genuine surprise on her face.

"Yes, really. We had this same conversation about ten years ago running supplies back and forth to the Jupiter moons. You always seem to forget the first college excursion to Mars when that young astronaught from the mines thought you'd played a practical joke on him. We got there in seconds and he didn't believe that you'd got the light alignment drive to work so quickly. He said you'd set the view-screen to an image of Mars. But we were actually there. Every time you tried to explain it he got mad and just said, 'Red flag!'"

"Oh yeah, that's right! Oh my, he was a very confused boy. That was over thirty years ago; nearly ten years more than how long Leena's been blinking her pretty little eyes." Yolanda gave a gentle but cheeky smile to Leena, who had just commenced construction on what could have been the biggest frown the Universe had ever seen.

Leena got out of her chair and walked over to hug the chief engineer from behind. "Yolanda, if I could live half as long as you and be anywhere near as wise I would die a happy woman!"

"Well, you have a long time to live, my dear, a long time."

Leena apparently had some cheek left in her as she asked, "Yolanda, given you are ancient..."

"Ow!" winced Emmalyne.

"That's okay, it's true it's true," Yolanda said as she waved her hand in the air.

Leena started again. "Yolanda, given you are ancient, what can you tell us about our upcoming destination, New Galapagos?"

"Well, only the basics; not like Hames, who actually spent a few months there. I only went there once and turned around pretty much straight away. We left Earth about a year after the Cataclysm on the only light alignment drive ship in operation. It was a great learning experience but a hard trip out, lots of reliability issues."

Hames raised his head. "That was the first time I met Yolanda, she took me out to New Galapagos – well, her ship did. It was an attempt to find out more about what had happened with the Cataclysm, to get some answers. It took about three years to get there; not too different from the pace we're doing now with lots of stops."

Yolanda looked at Hames and he looked down, not before he noticed Leena and Emmalyne catching the exchange of sad looks between him and the chief engineer. It had been a tough time for him, losing Nisha, and Yolanda had been a great friend to him on the trip out there.

"That sounds an okay place to start to me," said Leena. "I've been looking into the colonies a lot lately. I think I must have dozed off during Colony History at college!"

Emmalyne laughed. "Oh, I get frustrated going over the past; just ask Hames."

"You should try looking up Mermai, Emmalyne. I'm serious, it was really interesting to use the Williams Sphere to walk through my cousin's home planet."

"I might do that, thanks," Emmalyne said.

Yolanda continued. "Okay, well, my head is still sore from the accident but I'll do my best, and remember I'm no authority on it either. Um, when I went I was only in my early thirties. A junior engineer as I recall." The older woman was stretching out her injured arm, trying to energise it while she recalled events from years ago.

In Yolanda's pause, Dao Mai chimed in with, "I believe it may be a safe assumption to present to you all that Yolanda was on a ship which used the third version of light alignment drive technology to reach New Galapagos."

They all looked at him. His fact was a random one and their expressions of slight recognition and vague nodding seemed to be enough acknowledgement for Dao to return to his tablet and whatever was holding his interest.

"Yes, Dao, you are right, the third version tech." She looked at him incredulously. "I was an engineer on one of the old reconditioned light alignment drive ships. It was how I learnt about light speed travel." Yolanda clapped her hands together. "All right. New Galapagos was quite a small settlement. You have to remember these are basically big rocks sitting alone in space. Despite being far from the nearest sun they somehow maintain a capacity for life. The main planetoid New Galapagos is covered in clouds full of carbon dioxide and ice, so it looks like a proper planet from the outside. The clouds, its magnetic field and the solar winds that hit it made it habitable by dome. I guess the only reason humans are there is it can be reached by light alignment drive, like we're doing. They built a bio-tower on the tallest point of the planet so they could keep at bay all the microbes and bugs while pumping out a field that helps settle the atmosphere."

Xavi was getting excited. "I read somewhere about the sun-ray; have you seen it?"

Yolanda leaned back as if she had won a large sum of credit. "By far the coolest thing is the sun-ray. It's the first thing you'll notice as we come in. Up in the clouds it looks like a sun, big and round, but it ain't real. It's created using two beams. One oscillating laser from the other side of the planetoid shoots a circle-shaped containment field in the cloud above the citadel. Another beam comes from the bottom of the biotower to convert the cloud into the sun-like vision. It lights up the whole citadel, the biotower and the villages around it. At night, the beam is turned down so that the New Galapagos population can follow an Earth-like circadian rhythm."

"That's great," said Xavi. "A whole planet with an artificial sun."

"It wasn't always like that, though; before the bio-tower and the sun-ray, they lived in domes, like on the moon. When I got there the biotower was only recently built. Buildings were going up everywhere while commerce was blossoming. It was like they were creating a new place for themselves, a new identity. They had the start of a good lifestyle, plentiful resources and space to grow their lives and families. They were asking questions about what next. There seemed to be a gap in their purpose as a people, not a destructive gap, but a space to be filled."

Xavi iterated what most people in the mess hall were thinking and all knew; it was almost a group sense of pride. "They were looking for Universal Direction, yep." He looked at Leena, who gave him a pat on the back while she kept eating her food.

"You got it," confirmed Yolanda. "When we got there about four years after the Cataclysm, Universal Direction had moved forward with the people, really quickly. Not everyone was into it but the Administrators of New Galapagos were sponsoring it, which really helped.

"People told me when we got there that the first communications from Legeles de Nord created a real strong connection with the New Galapagos colony. After the Cataclysm they took it on for themselves in their own way. I'd only recently woken up to Uni-D and here was this amazing place that had truly embraced it.

Hames remembered this time. It had occurred as Yolanda had said. People had adopted Universal Direction in a frenzy. It was if they had been thirsty for water for decades and had only just found it. His first concerns about Universal Direction had been formed then in its early days.

Xavi was still beaming. "I love that story, Yolanda. Every time I hear the tale of how New Galapagos made a connection with the Universe I just feel so alive."

"Me too," said Chung. "I always look at it as the point which opened people up to so much more." Around the mess hall people nodded.

Suddenly, Hames did not feel like eating his last piece of salt cake. The others began a conversation about the first five years of Uni-D and what really helped people 'realise the truth they had always known'. Hames took his finished plate and cup to the utility wall and exited the room. No-one noticed and he felt a small pang of solitude at the non-recognition of his quiet protest against their beliefs. Conversely, he was happy to be let free from the meeting of minds and felt increasingly better as he walked away, the happy discussion in the mess hall dissipating and displaced by the hum of energy driving the *Medens* onwards.

Chapter 12

Sol-Concio

EMMALYNE WALKED GINGERLY AROUND the repair control room. The accident in the light alignment room had taken a toll and had further hindered her focus her time in stasis. Until this wasn't the case, Hames would watch his colleague closely to ensure her safety and that of the ship. A mistake in any of the recalculations with the repair engine had the potential for danger.

"The reminder's gone off, Chief. It's time to review the base calcs for the PCD repair assessments."

Hames rubbed the morning from his eyes for the fourth time that day. "Okay, the Lambda-CDM file should come up ready for execution."

"Chief, how many times have you had to do this, check the core levels of dark matter, I mean?"

"Too many. We have to assess what needs repairing to know what is left over."

"I guess so." Emmalyne continued at her workstation. "I did look back at your old tests. Every month for the last couple of years there has been no change."

"Protocol you know, and it's not like we're busy," suggested the chief mission officer.

"Sure, why not." Emmalyne turned to one of the screens nearby and started working away. "I am now undertaking a fruitless check with the LCDM model."

"Thank you." Hames gave Emmalyne a mock half-bow to welcome her initiative.

"How far ahead is the forward scanner reaching?" he asked Emmalyne. The young mission officer was moving to a large view screen on the right-hand side of the repair control room. The viewing space showed the immediate path of the *Medens* through what looked like normal space. Nearby, a smaller terminal produced a holographic three-dimensional view of what appeared to be a light tunnel piercing their path into the debris of the shattered conversion trails.

"It's shooting ahead about four clicks. We've more than enough information to see what is coming."

"That's great. Let's finish mapping the next stage and see if Francois can bring us something to eat. If we work through this last check we can finish up early. I reckon that'd be good for both of us."

The repair stage didn't take long to lock in. There were some points of degradation in the debris trail, a variance that had to be accounted for. Throughout this process Hames would grumble to himself, scrunch his greying beard, before executing a final tapping of his hand on the workstation; not loudly, more in slow motion as if done subconsciously. "We're not really making the best time at this stage. I hoped we'd be further along."

The young mission officer turned to face her boss, her voice assertive. "Chief, we're never going to heal it all. Well, not for a couple of decades. Not unless we figure out a quicker way. It's like you actually think we might make it to Mermai."

Emmalyne appeared annoyed. Hames recognised the mixed messages he was giving her. His frustration about not making good time on mending the fractures was against the practicalities of the science of repair. However, the impossibility of the task was always overridden by the desire to reach Mermai. To even be at a point at which they could communicate with Mermai would be amazing. To speak to anyone in the Mermai system who could tell him if she had made it there. Did she have a good and happy life? Did she find someone to love? Did she wonder what had happened to him? The possibilities of these answers drove him onwards each and every day.

Francois entered the room carrying food and drinks, smiling and content. To Hames' surprise he was followed by Olisa. Pleasantries were exchanged and the old healer sat down next to him. She turned to Emmalyne across the room. "How are you today, my dear?"

"Fine thanks. It's been a busy morning." Emmalyne eased herself away from her workstation, wiping her hands down her sides. "I'm starving. Francois, what have you got for us?"

The tray placed in front of both Emmalyne and Hames revealed a Mermain feast of dried meats, chocolate and mint salads.

"I love Mermain food; I adore their version of mint. I will eat it with anything – chocolate, rice, bread, it goes with anything!" Emmalyne looked as if she was drooling.

Olisa looked somewhat appalled but offered agreement. "Yes dear, Mermain mint is, ar, special. The COO's fondness for it is quite well known."

Emmalyne piled a plate high with salad and meats before returning to the waist-high bench top that held screens and tablets. "While we break I'm going to read up about Sol-Concio. Leena's into all this history and the Chief reminded me about it the other day in the medical-pod; it got me thinking."

Hames was impressed and called over, "I'm keen to know what you think goes on out there." In all respects, the inquisitiveness of his young charge should be encouraged, certainly not impeded.

Olisa also seemed to be thinking along the same path, only to present it in her favour. "It's a fascinating thing that you are reading up about it. I'm interested in your curiosity. Only a strong enquiring mind would do such a thing."

"Thanks," responded Emmalyne.

Francois moved up next to Emmalyne. "Can I watch it with you? I always seem to have more questions then answers when I search on Sol-Concio."

"Sure, take a spot." Emmalyne shifted her tall frame sideways as she leant her belly into the bench, arching her back and eating with one hand while the other flicked through the screens in front of them.

Hames could not avoid noticing that the frame of the large Cernunnosian obscured Emmalyne from his view. This gave him a sense that the two groups had been segregated. For now, he was in a one-on-one situation with Olisa, something he tried to avoid on a regular basis, even more so since the healer's pursuit of Emmalyne had begun.

At her desk, Emmalyne clicked and typed data into a clear screen interface. Out came a projection of the Milky Way. She tapped a few more times on the table and this view began to focus on the giant Orion Nebula, its unmistakable pink and orange tones suffocating a sea of stars. She then pressed a button to start the analysis of Sol-Concio:

Sitting hundreds of light years from Earth, Sol-Concio is the third furthest of the seven PCD colonies. It is the second planet of a group of eigh planets, each with moons, all spread a relatively equal distant apart. The system sits in front of the lower parts of Messier 42, the orange, red and yellow tinged nebula, otherwise known as the Orion Nebula. Although this group of planets sits a long distance from its two suns, the proximity to the nebula that towers over the mini-system is said to provide it with an atmosphere that is consistent, warm and predictable in that it does not have seasons such as other planets. This atmosphere drawn from the nebula was the critical factor in identifying Sol-Concio as a suitable colony and regular PCD route.

"Wow," said Francois, "I never knew that."

Emmalyne paused the viewer. "It's easy to see why it was considered a special and sought after place. Can you imagine being on Earth and having the same temperature all the time? It would be like being on holiday every day." Her high cheek bones rose and her eyes twinkled at this idea.

Hames could not help but join in. "Some people believed it could be like heaven. Hanging there in the stars in front of this magical cloud that is light years long and wide. It was one of the reasons the Corona Movement fought so hard to settle there."

Olisa, for all her positivity and open-mindedness, appeared to dislike this idea; her face had been growing longer throughout the conversation. "There is no heaven, Chief Hames, no afterlife. The life we have now is part of all life; we are everything, as Universal Direction tells us."

"Oh, I don't know," said Hames, "maybe I'll die and resurrect myself as a ray of light beaming down on Sol-Concio." Now Hames was being mischievous. He despised being told how to think by Uni-D and struggled not to show his disdain. This challenge to look happy and calm was made all the harder by Olisa focusing all her attention on Hames with her words and body language.

"There is no death, no birth. Just the ever-presence of you, and I, in the Universe."

"Okay, okay," Hames held his hands up in defence. "You got me. I don't believe in heaven, I don't believe we're all connected beings either."

Olisa looked at Hames, dropped her head and exhaled. "You may not recognise it or acknowledge it; nevertheless, we are meant to be here together. You and I are on this ship together."

"With respect, that is somewhat ridiculous and for the main part unprovable. Firstly, I'd never met you til a couple of years ago. Secondly, you and I, we see things very differently." He struggled to look her in the eyes and pretended to play with some mapping, hopeful that she would go away.

"Chief, even Albert Einstein, a scientific mind like yourself, suggested that a human is part of the Universe but that we, as humans, are limited in time and space. Einstein considered time and matter to be a kind of optical delusion for our consciousness. Don't you agree?"

"With Albert? Most of the time; there's a couple of things he got wrong so no-one's infallible."

"One day you will see we are connected. One day you will turn around and thank me."

Her statement made his blood boil as he reached his point of tolerance. He felt his face burn. He looked down at his hot chocolate, recently brought to him by the ever-generous Francois. Faint clear wisps of steam sought escape from the edge of the cup, like the ghost of a small flame that had never existed. Olisa waited for him to speak. Her grey overalls looked baggier than usual, hanging about her like a cloak, and her sleeves were long, covering most of her hands.

Hames began to lose his appearance of communal patience as his mind pondered the reason for this interaction. "Why are you here, Olisa?"

"That is a very good question, Chief Hames. Why are you here?" Olisa's tone was just as calm despite her loaded question.

Hames was never going to answer this question and not because of his inherent disregard for its premise. He was confident about the purpose of the healer's visit. "You want to practise on her." Hames was getting a grip on his emotions. "Like I've said before, she doesn't need fixing, she's fine as she is."

"And who is to say she doesn't want the experience? She has been studying for many years, readying herself," countered Olisa. "She doesn't need to be broken to agree. From Curate-mente she could simply be improved."

"I don't like the sound of that. By improved you mean be more like you, more staid and more, um, reduced. I've seen people retreat into themselves so much that they cut themselves off from the world!" The loudness of voice caused him to check himself. The pause to this civilised bickering allowed them to listen in on Emmalyne and Francois' study on Sol-Concio.

> Following identification of the seven original PCD routes, all of the end points were marked out for like-minded communities and nations of Earth to extend to. All of Earth contributed to the settlement in establishment of the new colonies. However, Sol-Concio stood alone in that it was funded and settled completely out of the resources of the Corona Movement. The Corona Movement was one of the non-State aligned entities or Free Factions that grew out of the Freedom Act of 2242. The Corona Movement is also notable for its part responsibility in the sponsoring and introduction of Universal Direction by the predecessor to the Organisation, the Interstellar Agency. Sol-Concio was named with an intention to refer to the Sun-Mind or Sun-Consciousness as the Corona Movement had a focus on the impact of solar rays on humans bodies, minds and spirits. The settling of Sol-Concio solely under the auspices of the Corona Movement marked it out as the only 'private system' in the Universe and this assisted in establishing a short-term pattern of selective migration and interaction.

"It must be a special place to have received all that attention." Francois wiped his large dark hands over his belly after finishing some cured meat.

"I know, it's pretty fascinating isn't it? It's like they had a big secret they didn't want to share or a special club that only special people could join." Emmalyne emphasised the word special with a smile.

Olisa patted the workbench in front of Hames to gain his attention. "Whether she chooses or not, the girl definitely needs healing. She's wobbling about like her legs are jelly yet there's physically nothing wrong with her."

"There is nothing wrong with her, so she doesn't need to be healing" Hames was shouting in a whisper and looked over at Francois and

Emmalyne to make sure they hadn't been heard. Emmalyne turned up the volume on the viewer:

> By comparison to other colonies, communication with Sol-Concio was challenging and inconsistent due to the small number of craft allowed to journey over the Sol-Concio Tentacle. Despite Cernunnos and Lugus being further away from Earth, regular travel to these destinations provided much more information as to the development of their colonies. Little is known of the internal governance of Sol-Concio nor the daily lives and activities of its citizens.

The voice from the old healer took on more certainty. "By taking Curate-mente, Emmalyne can take herself towards a greater state of being. She can reach beyond the self to see the all."

"She doesn't need to take on mind-healing to reach a greater state. Just look at how far we've all come in the last few hundred years. No disease, poverty is gone, we've travelled across the galaxy and back...well, the last one hasn't worked out so well, but my point still stands. Evolution is all we have needed to reach this point."

Olisa's arms started to circle in an expressive fashion. "Evolution is a myth. It's just a physical construct, a tool of the material world. It perverts and prevents any realisation of what we truly are. There was no genesis of life. The genesis of life is a lie to stop you from seeing that we come from the field, from the source, from the Akasha."

"The Akashic Field? I've heard this too many times and not once does it ring true to me. One big field of energy that we can all somehow tap into. Show me one good example of this, just one!"

"That is Curate-mente, it helps us touch the field. If you tried it, you may understand that to be the case. You refuse so you cannot understand. You limit yourself and that is your choice."

Olisa continued while Francois and Emmalyne viewed the analysis. "I question your judgement in these matters. You are using your own proclivities as a basis for making a decision on how to treat young Emmalyne."

"My proclivities, my proclivities! So that's what this is about. You say you're understanding and preach tolerance where as you've no tolerance for me not aligning to your beliefs. Also, you haven't a clue what my proclivities are! You don't know me and you're never going to turn my mind to worm soup trying to find out!" Hames snapped at Olisa, who looked back at him and shook her head. The healer pushed away from him and moved away.

Emmalyne turned to face Olisa, who was now selecting food from the Mermain feast brought by Francois. "Look at the Orion group system here, the length of the nebula. It's massive – light years long."

"Do you like the nebula? I think it's beautiful," replied Olisa. Hames paid no notice to her. To him Olisa was more faith healer than medical practitioner. He had heard the expression "witch doctor" before and he was sure it applied to her. His silent consideration of this name for her was punctuated by further analysis of the distant colony.

> It is rumoured that eight years prior to the Cataclysm an armed secessionist dispute between two factions of the Corona Movement occurred. Limited intelligence is known of the dispute. One group favoured opening up the colony to the rest of the Earth network while the other sought minimal visitation from outside colonies. The assumption made by offices of the Organisation was that the group seeking continuation of the private status of Sol-Concio won this dispute. This assumption is supported by the ongoing and limited communications from Sol-Concio after this period. Further, from 2365 the amount of travellers from Sol-Concio decreased even further, suggesting a number of ships had departed this colony permanently.

Emmalyne turned to her right and took the two hands of the person watching her read the analysis. "Mistress Olisa, it has been highly beneficial to have your time of late. Already I feel better just talking to someone such as yourself. I feel I'm starting to shake off the feelings from stasis. Maybe that accident in the turret observation somehow helped."

"Very well, girl, very well. If you feel a change for the worse just come and see me." With this final message delivered, Olisa and Francois left the room.

Hames joined Emmalyne and they finished studying the Sol-Concio images.

"It's peaceful and calming, the nebula, isn't it?" Hames leaned in with her, both of them with their elbows on the table, hands supporting their faces.

"I always think it looks like such a piece of interstellar art. I just wonder whose work?" Emmalyne propositioned.

"Maybe it's a mystery we're not meant to solve. Maybe that is part of the beauty of nebulas; if we did know everything about them it would spoil their beauty."

They sat there as the images of Sol-Concio ran through the viewer. Hames then moved away and back to the programming of the tentacle data for the repair engine. He cast a sideways glance and half-smile over at Emmalyne. She saw this and then motioned to the screen playing the analysis of Sol-Concio. "Do I need to finish up with this?"

"No, not at all. I wasn't even really looking at you, just thinking. Keep going, it's not a problem."

"That's alright, Chief, I think I'm done anyway. I can look another time if I'm interested. Let's see if we can't finish for the day." Her smile back at him was honest and hopeful. She portrayed real and present emotions that he could not in any way conceive as out of place. Here was a person with a true open spirit. The possibility of Olisa's "healing" scraping this out from Emmalyne's core, or even ameliorating any part of her just one little bit, filled him with a terrible sadness.

Hames was about to raise his concerns for her future when suddenly, the metal cup on the repair desk vibrated – just for a second and then it stopped. A hum began to sound out in the room, moving to different sides randomly. It was like a fly or bug flitting around him. The hum got louder and the sides of the round room began a random jerking in tune with the hum. Then, a short silence preceded the hum's ascent to a noise that began to penetrate his bones. Hames felt a moment of fear. He had no idea what was happening to the ship. A second later the ship emitted a pained cry of metal and plastic all around them. This was a scream that told him the *Medens* was about to tear itself apart. He stood up and moved to the repair engine monitor. A wave of panic hit him when he saw the readings on the repair engine showing a positive output, which should never be the case.

The repair engine was malfunctioning, disastrously so, and making the early preparations to send the *Medens* off on a new pat h. A path devoid of him, his friends and any hope of finding Nisha again one day. The *Medens* was about to fail.

Chapter 13

Particle Overflow

A LOUD AND PERVASIVE ALARM RANG throughout the *Medens*. On the gangways and in the mess hall. Through the home-pods to the bridge. A constant emergency signal that deviated up and down a C-octave, back and forth. All over the ship, heads turned and work ceased. The lighting strip that channelled itself around the top third of the *Meden*'s corridors strobed a sporadic emergency notification that moved from ambient yellow to a blazing purple; this signalled to all that the danger was authentic.

In the repair control room, Hames felt fear flow through him and then dissipate as he realised action was required. In his mind he pinched himself and he started a run towards the exit of the control room. He heard Emmalyne shout an obscenity behind him. She too must have seen the reading on the monitoring screen, one that showed a positive output in the repair engine. Having Emmalyne attuned to the imminent disaster was a small positive note to the situation at hand.

The ship shuddered violently as he ran down the corridor, shoving him to the left and then down into the corner of the corridor wall. Though not overly stunned his attempt to stand was defeated by a small wave of weight that stretched into his chest and down to his legs. This wave of heaviness was enough to confuse his body and tumble him back onto the hard plastic floor of the gangway.

Emmalyne, who was following close behind, did the same, falling into the space beside him. "Chief, it's gone! It's like the repair engine has gone from the *Medens*!"

"I know! It's broken somehow!" Hames shouted over the back of his shoulder. "It's retaining particle debris and showing a positive reading!"

He tried to pull himself up but again fell down under the heaviness of his body.

"What are you talking about?" Emmalyne shouted back.

"The repair engine, it's in a positive reading status. It can't do that. It means we're not pushing the converted particle debris out of the ship, at least not quick enough. I have to get to the Engine Unit to see what's happening!"

"When I looked there was nothing! It showed no repair engine on the *Medens* at all. It's like it vanished from our system!"

Hames realised then that the incident was only just beginning. "Argh, even worse!" Again he tried to move upwards and forwards but another feeling of weight threw him off balance and back to the ground. "The *Medens* is in terrible danger. The debris could fill up the ship and combust it. We've got to get to the repair engine now!"

"Come on then, you old dinosaur!" yelled Emmalyne over the vibrations and alarms. "It's just the gravity well playing tricks on you. I reckon it's being maxed out by what's going on with the repair engine!"

Of course, he should have thought of it. They pulled themselves up to a standing position, leaning on the wall. With the knowledge of why their new-found heaviness was occurring they could adjust their movement to ensure onwards movement to the repair engine.

"We're getting heavier!" Emmalyne shouted back to him as they moved further along the corridor.

Hames knew she was right. He looked over his shoulder with one eye closed, trying to adjust for the beaming lights. "The gravity well, it's not regulating itself. The repair engine is pushing extra power out across the ship. To the well and to the particle collector. The whole ship is going to become super-charged and pull itself to bits."

"Well, what are we going to do about it?" Emmalyne's hazel eyes were also squinting as her blonde hair fell out in strands from its tight ponytail.

Hames stopped. He tried to contact Yolanda and the COO on his communicator without success. He tried the ship's first pilot. "Leena, Leena can you hear me?"

"Yes, I hear you! What have you done to our ship, Chief?" Leena shouted out her barely audible question to him through the communicator.

"Enough of that, Leena! We need to drop out of light alignment mode and punch the cruising engines hard, really hard!"

"Why?" responded Leena.

"We need the cruising engine to use up our energy and slow down the consumption of the repair engine!"

"What?" screamed Leena.

"Just do it!" Emmalyne screamed to Leena through the communicator in her brown wristband. She then put her hands on Hames' shoulders. "Let's get on with it then!"

They began the slow movement towards the repair engine room. The ship was vibrating so much that Hames started to feel nauseous. It was taking forever to make the short distance to the room from where their current predicament was emanating. A distance that would normally take a thoughtless two-minute walk now required every effort of muscle and sinew to attempt it.

Finally, they reached the entrance point to the repair engine room and the ship performed a massive shudder. Leena had done it. The light alignment drive had moved offline and the ship was moving into cruise mode. Following this latest shudder, Hames again fell forward and onto the floor of the *Medens* gangway. This time it was from lack of weight and the force of strength he had been putting into his forward movement. Emmalyne caught herself on the side of the *Medens* and strolled gracefully into a normal walking movement. Once she truly caught her balance this turned into a jog through the entrance point of the repair engine room. For a moment Hames marvelled at her athletic ability and grace of movement in such a high pressure situation. However, he couldn't dwell on such thoughts while there was incredible danger to him and his crew.

Pulling himself up and into the repair engine room he could see Emmalyne checking over Chen Chung, who was leaning against the top of the stairwell that led down to the repair engine interface block. Chung had a large cut running across his forehead that was beginning to bruise and bleed. Despite the gravity well approaching a normal level, Hames was still feeling seasick as he approached him. "What's going on here?"

Chung raised his hand to hold his head. "The collector is taking in more debris than the ship can handle. It's sucking it in, not pushing it back out."

Emmalyne held Chung up from the stairwell rail. "What's Xavi doing?"

"He's trying to trigger a manual release of the debris from the repair engine, to get it out to space. Something's caused the engine to close up. We don't know what though." Chung was exhausted and exasperated. "We just have to get through a lot of other systems before he can do this."

Hames looked down at Xavi working around the large grey, rectangular interface block. The immense shape hummed and glowed deeper than he had ever seen before. He turned back to Chung. "What are you doing up here?"

"I've been trying to get the interface block to link back in with the ship. It's moved completely out of the *Medens* systems although it's still clearly there." Chung gestured with weary hands down to the large rectangular shape below.

Emmalyne shouted over the alarms. "We know! I saw it had gone when we checked back in the repair control room."

Chung yelled back, "I came up here to the workstation to shift the unit back into the ship's systems, but I fell over when the gravity well came back online."

Hames pulled them both in close to him. "Okay, let's get this sorted. We'll tackle it on two fronts. Emmalyne and I will relink the interface block back into the ship so we can give it a command to release the debris. Chung, you help Xavi activate the manual release point.

"Where do you want me on the relinking, Chief? Up here or down at the block?" asked Emmalyne.

"You take the screen on the interface block. I'll take the systems up here." Hames went to work on trying to re-establish the interface block in the ship while Emmalyne made her way downstairs. Chung moved quickly back to Xavi at the far end of the large device to begin their work on the manual intervention.

After pulling down a data point, Emmalyne began to work furiously at the end furthest away from Xavi and Chung. She looked up, calling out to Hames, "The repair engine is taking in too much debris; we need to get to the collector to make it release!" Emmalyne frustratedly punched in recalculations of the various stages they had mapped. Her fingers were like a piano player's, long, sleek and strong. Under the vast pressure that was the ship's base structure seeking to tear itself apart, she worked intensely. "It's like the whole Mermai Tentacle has been doubled. The repair engine thinks it's venting the right amount but there is loads more in there!"

"Can you sort it?" Hames yelled.

"Every time I do something it resets. It's not working, Chief!"

"We've got to slow the volume of particle debris. Get the collector turned off and move the ship towards a section of space without any debris."

"That'll take forever!" yelled Emmalyne.

"Just get Leena to start taking a deviation. It will eventually help the repair engine ease off."

Emmalyne pressed the eight-circle symbol on her leather wristband. "Leena, Leena can you respond?" Nothing came back. "Leena, Leena what's happening?"

A crackle came back. "Yes what?"

"Move the ship away from the debris!"

"Really?"

"Yes. Chief said so!"

The change in direction pulled the *Medens* away from the particle control debris but also seemed to distort the environment around them.

"Chief, I'm starting to feel crook," Emmalyne called.

Hames looked down at Xavi and Chung, who both looked affected. He was doing his best to focus on the screen in front of him and then down to Emmalyne for her signal. But the alarms, vibrations and the violent shuddering made him feel nauseous and as if he was moving through treacle. Completely disorientated, he focused on the screen and rebalanced. He tried another combination of codes and links, then looked at Emmalyne; she appeared to be struggling against the same feelings. She leaned forward on the end of the rectangular unit, resting her head on her forearms. He could see in the background, Xavi and Chung moving about the floor with slow steps. Back and forth they went for tools to get the repair engine to trigger the particle debris vent at the bottom of the ship to open up.

Emmalyne was working harder and harder on her code to re-establish the line of connection with the ship. Chung was half a body deep into the end of the interface block, handing out pieces to Xavi, when there was another almighty shudder of the *Medens*; a large crack sounded somewhere beneath them. The interface block shifted and with this modification Chung's body snapped and went limp.

"No! Mind no!" screamed Xavi.

Through the treacle-like haze, Hames could see Xavi's distress at his friend's end. The repair engine officer was in shock and backed away from the sight of Chen Chung's body, limp and hanging out the side of the interface block. Xavi's hands rested on his head and then slid down to cover his face, locked in his pain. At the same time, both Hames and Emmalyne saw the next danger. Emmalyne began shouting out to Xavi. Shouting to warn him of the creeping field of black matter that sparkled dully and reached outwards from the opened end of the interface block. Hames himself tried to warn Xavi, but he was powerless in the noise and calamity.

The particle debris field touched Xavi and a flicker of energy visibly passed through him, balancing Xavi's field to that of the debris. Energy pulsed throughout him, although it did not illuminate him. Rather it produced what at first looked like shadows on and around him. Xavi dropped his hands with the realisation that he too was in a situation of fatal consequence. The look of fear that both Hames and Emmalyne had to witness was one the chief mission officer would never forget. For a second, Xavi's terrified outline stayed the same while his insides were converted to the particle debris matter he had so desperately been trying to eject from his ship a mere moment ago. Then the ship shuddered, pushing the previous form of Xavi outwards to something unrecognisable.

Hames tried one last combination then looked to Emmalyne. Finally, the interface block was live in the *Medens* systems again. Somewhere beneath the ship a vent opened and the particle debris loose in the repair control room began to move back towards the interface block. Hames, in his foggy daze, could only hope that it flowed out of the *Medens* all together. The ship vibrated once more and he could feel the gravity well resetting while the last of the particle debris vanished into the interface block.

Hames made his way down the stairs and joined Emmalyne at the end of the large rectangular device, which hummed and glowed commensurate with its previous parameters. He stood back cautiousl as Emmalyne fell to her knees a short distance away. Then, in silence, Hames circled the large device.

Tools and parts lay everywhere. Chung's body still hung from the other end of the block. Emmalyne stood to join Hames in his inspection. As they drew close to the still open end of the device a horrifying moment occurred. The shape of a human in particle conversion form surged forward out of the unit, not at them, rather up and outwards with the end of its form dragging behind. The shape produced a noise. The outline seemed to shout in his mind the words 'no' and 'light' at the same time. In his body and mind, Hames felt the last words of Xavi. As quickly as it had appeared the shape dissipated back into the interface block.

Finally, with shock on his face, Hames went over and pulled the body of Chen Chung away from the interface block and shut the opening that Xavi and Chung had been working in so feverishly to save the *Medens*.

Silence embraced them before an almighty wrenching sound of clunking metal told a story. The *Medens* had come to a complete stop. They both jolted forward. Hames flew sideways across the room. By the time he hit the ground he was out cold, unaware of the damage and devastation all around him.

Chapter 14

Battersea Park

NISHA WALKED THROUGH THE corridor to exit her building and experienced the theme of that week. Stifling heat wafted all around her as she trudged through desert sand, blinking against the brightly lit skies. On the walls, she saw thousands of slaves pushing heavy blocks into pyramid structures. There was just enough movement in the diorama to bring it to life. "They have really outdone themselves this time," she spoke out loud as a group of men garbed in loincloths pulled a final triangular stone atop of the famous construct. By the time she had reached the end of her corridor the radiant heat emanating from the walls had begun to make her sweat.

She welcomed the morning breeze that blew down Battersea Park Road. She left her building feeling refreshed and reacquainted with the temperate English summer day. This momentary feeling of ease was interrupted by the knowledge that her short walk to the doctor would be monitored by someone from the Agency.

Nisha reached to her neck for a point of comfort. Earlier that morning she had rummaged through her storage to find an old leather necklace that dangled a Sovann Macha pendant. The image of the forlorn mermaid reminded her of her home and its cool metal against her olive skin made the connection feel more tangible. Her homesickness had grown lately. Whether it was the incident at St Paul's or her approaching return date to Mermai it didn't matter. Earth was starting to feel an unwelcoming location.

A small bud communicator nestled in her ear delivered Hames' concern about her wellbeing. "I want you to take care of yourself!" he pleaded.

The device carried the love of her life in her head but it was slightly pestering her. "I'm doing that, you silly boy!"

"I love it when you call me silly. No-one else does that here."

"That's because your bosses are too busy kissing the arse of the star student! Or is it the student of the stars, I get confused." Nisha laughed as she turned a corner into another street.

"That happens all too easily. Where are you?"

"I've left home and I'm walking up towards this new doctor's place. The Agency are still here Hames, watching me." Nisha scanned the street around her to see it was empty. "I don't know where they are, but I can feel them."

"I'm sorry; there's not much I can do about it – I tried."

"I know. It's so just undeserved, so undeserved!" Nisha slapped her olive-green hand on a low brick wall as she past, tossing her deep brown hair indignantly.

"Just try to ignore them. They're also there for your protection."

"I don't like it. I really don't."

"I know. Bear with it and let me know how the doctor goes. I won't be home late, I promise."

"Sure, sure, see you later." Nisha closed down the communication line and continued on Queenstown Road to find the new doctor's surgery. Nisha looked up. The sun glistened off the tall silver and glass buildings that dominated the skyline of south-west London. In reflex, she directed her blinded eyes down to the street level and the restored red-brown brick buildings. They appeared old, although she understood them to be relatively new. If not for the sound of hover-transports above, she could have believed herself to be in the London of centuries ago.

When she reached her destination it was clear that the building was not purpose-built. It looked more akin to a shop entrance with its large front windows. There was nothing to declare it a venue for medical practice, simply a sign that read 'Open' on a display screen affixed to a window. She proceeded nervously, unsure if this was the right place. Behind a counter sat an old man in a white moulded chair, positioned in front of an entrance point to a back room. Against the far wall sat two small tables with viewing screens set within them. Despite the shabby and uninviting nature of the exterior the whole ambience was pristine, white and bright.

"Welcome to the Battersea Universal Direction Centre," said the old man, coming to a standing position. His voice was cracked and his body frail. The smile, though, was one of the most genuine that Nisha had ever seen. It pleasantly distracted her until the realisation of what this location was dawned upon her.

"Oh my, this is what the announcement in St Paul's was about, and a flyer I was given in front of Clapham Junction. Something about harmony for Earth, being together forever?"

"That's right. This is a Universal Direction centre, for healing."

Nisha backed away and half opened the door onto Queenstown Road. "Oh, no way. It's caused me no end of trouble. I was there when the bomb went off!"

"Well, then this is the perfect place for you. You seem sore and bruised." The man's smile broadened in a cheeky manner. "But not that bruised."

Nisha instantly relaxed at the offering of humour, feeling her tense shoulders drop. The old man now sounded fairly normal to her. "Okay, so you're the doctor then?" She closed the door and walked back into the healing centre.

The old man laughed. "You have me there; I'm not the doctor. You're not limping either so I just wondered if you were here for something more than your bruises and bumps?" He looked back at Nisha enquiringly.

"I don't know," Nisha mused. "Mainly I'm just feeling really sore and I'm looking for some shortcuts to recovery."

The old man put his hand to his mouth and was thinking about something. "Why don't you take a seat and let the healer see you. As I said, I'm not the doctor so we'll see what he has to say."

This was not what Nisha believed she would be doing today. It seemed questionable as to whether there was a doctor in this building or anyone who was going to actually examine her shoulder. Unfortunately, it would have to do as Nisha's pain was at a level of discomfort to trigger grumpiness.

Hames was always telling her that curiosity is a good thing. So perhaps she would stay for a short time to see what it was all about. "Fair enough. Shall I take a seat? I see there's a couple available." Nisha grinned while she motioned to the white table and chairs.

"Anywhere you like," the old man responded, waving his hand around the room. "Healer Milar will be with you soon."

Somewhat tired, Nisha moved gratefully to a seat and then realised this was not the normal routine of visiting a medical practioner. "Hey, aren't you going to ask permission to link to my profile? You haven't linked to it without permission have you? If you have, I'm out that door straight away!"

"Healer Milar doesn't need to see your profile; he just needs to see you."

"What if I have some sort of ailment that he needs to be aware of? If I'm colour blind or have a history of back pain?"

"What you will do with Healer Milar does not require any of this knowledge." The old man again gave Nisha that lovely genuine smile. She tried to look through it to find a hint of patronising insincerity, but for all her efforts could not find a drop.

"You're very into this stuff. Aren't you a bit old to be adopting such a radical idea?"

"It's not radical. I've had belief in harmony all my life. I'm just happy to finally start helping others see the world as I see it."

With a wide-eyed nod Nisha acknowledged the views of the strangely committed man and neither outwardly countered nor supported his belief.

She eased herself into the white chair that moulded itself around her. The old man himself was still grinning. She decided this bordered on making her feel uncomfortable so she looked down and placed her hands on the table. At her touch, a screen started shifting within the plastic table towards her, setting itself in front of her. Up came an interface link titled *The Hammadi Herald*, which looked like a news service. She clicked on it and a welcome screen came up:

> Welcome to the Hammadi Herald. *This is our first edition and we ask that you celebrate the information we are about to bring you as part of the commencement of Universal Direction, announced by Legeles de Nord...*

After the welcome note Nisha worked through some of the other parts of the document. The front page was a report on the announcement by Legeles de Nord in the square. It outlined the key conventions of Universal Direction and listed all the people and countries around the world adopting it. It wasn't until she scrolled down a few pages that she came across a reference to the bombing she had so painfully been a part of. Why was it so hard to find? It was almost like the editors of this publication were trying to hide a massively significant event from anyone seeking to commit to this new secular approach to life.

She scanned the document and found large adverts for new Universal Direction centres opening up all over London and around the world. Further, it all appeared to be at no cost to those who visited these centres. No credit was required and the message was, "don't come once, come often". Services ranged from meditation, light education and Curate-mente, whatever that was. There was even a section on recruitment for healers and administrators of the new centres.

She looked up at the old man sitting back on the white chair looking so peaceful. If she ever found the shuttle transport hub too much of a burden maybe she could have a career change. The idea of sitting on a chair all day, talking about the harmony of life, did have some appeal.

Finally, the door behind the old man opened up and a small slim man in soft white pants and a white three-quarter–sleeved shirt came over to her. His wavy mousy-brown hair was not too long and some flecks of grey suggested his age despite his youthful smooth skin. His square jaw bore no weight and neither did the rest of his trim body. What struck Nisha most was the calm and friendly manner emanating from his relaxed walk towards her.

"Nisha, pleased to meet you. I'm Healer Milar."

Uncertainty filled her normally confident voice. "Thanks, I'm not really sure though if this is the right place for me."

"I can understand that. If you'd like to come through you can get a better idea of what we can offer you. That may help you decide whether this is the right place or not."

Although she did not want to acknowledge it, his presence made her feel at ease. "Sure, I've come this far." Nisha then followed the healer into the back area of the strange-looking medical centre.

What lay in front of her appeared more than a clinic. White cupboards and chairs sat back against the walls of the room. Its central piece was a square mat that covered at least three-quarters of the floor space. Within this area were four curved mats in an oval shape, all of equal size and distance within the square. On one of the mats a man lay perfectly still, staring up to the sky through a window above him. From the other side of the room Nisha was unable to tell if he was sleeping or awake. It all seemed so strange and she felt her nagging discomfort increase. Reaching through her long brown hair she grabbed the Sovann Macha pendant and held the metallic mermaid tightly.

Healer Milar invited her to join him to the left of the large mat where two white stools awaited them. Coming to a rest on the stool the healer brushed the top of his legs as if they were dirty. He then pointed to the centre of the room.

"That's my patient, Christopher."

"Is he okay? He doesn't seem to be moving."

"Yes, perfectly fine now; he's been with me for eight months. Been through a lot in his life. Christopher Harkness used to be fairly notorious; he was a leader in one of the feudal gangs from south-west England. Quite violent and depraved. You can look him up if you're fascinated," offered Healer Milar.

The man lying down was of a normal build although his face was weathered as if it had seen too much of life. Various small scars around his cheeks and neck enhanced the appearance of degradation; lank black hair framing his still image. Healer Milar spoke very plainly about the man, as if he wasn't there at all. "He was robbed of his empathy as a boy. He lost his mother before he was one and his father was incredibly violent in that devastation. Christopher spent much of his life in and out of internal service, until he was tasked into mine."

"What do you mean? Does he work for you?"

"No, Christopher was given a choice. Heal or lose himself to the state forever. He chose life, he chose harmony."

Nisha had that weird feeling of being dislocated in her day. Right now she should be receiving some painkillers or a massage. Instead, she was reviewing the re-education success of a well-known assailant. She stared at the calm man on his back with great caution.

Healer Milar changed the subject. "Have you heard much about Universal Direction?"

"No. Well, yes I have heard what it's about. I can't say my first interaction with it was very positive."

"How do you mean?"

"Well, when the bomb went off in front of St Paul's the other day I was there."

"Oh, I see."

"Yes. It kind of beat me around, which is why I've gone looking for a doctor." Absent-mindedly Nisha pushed her hand inside her small dark navy jacket to rub her left shoulder.

"Some people are afraid of Universal Direction, what it may bring to Earth."

"Should I be afraid? The bomb explosion was frightening." Nisha had pondered this question much, especially in the first days after St Paul's.

"This fear, fear of change. Doesn't that tell you that something special might be happening?" asked the healer.

"Um, no, it tells me that not everyone wants Universal Direction. It looks like it's creating disharmony, which is ironic as it's trying to create the opposite."

"Universal Direction is the policy of the government of the day. They were elected to set the strategy for society. This is their strategic plan." Healer Milar gestured to the space around them. "To reject Universal Direction is to reject the strategy for society, set by our elected officials."

"What if there is no harmony in propagating Universal Direction? I feel uncomfortable when there is conflict around me." Nisha held her pendant tightly. "On Mermai we all generally get along."

Healer Milar spoke in an even quieter more comforting tone than before. "Mermai, do you miss it?"

"Terribly. It is my place with my friends and family. It's full of rich colours while all the colours on Earth just meld into one big splodge to me."

Milar laughed. "You're right, we can certainly look a bit muddy at times." His small smile faded as he asked the next question. "Do you think it coincidence that you were at the explosion then found yourself here?"

"Maybe, I don't know. I never considered it that way. I really just wanted to see a doctor, that's all."

"To heal yourself." He pointed to her arm.

"That's right, to get better and hopefully quicker than what it seems to be taking."

"Do you remember what was said at the gathering before the explosion?"

Nisha was suddenly aware that the healer was taking her places in the conversation that she was certain were unrelated to healing her shoulder. However, she felt it would be polite and harmless to play along with his ideas. "I remember most of it. Afterwards I read lots in the news streams about the gathering. I also saw some info on a pamphlet promoting Universal Direction. It was about meditating, healing and the Earth becoming harmonious again.

"That sounds about right to me. The part of the healing, though, is different to what you might normally expect from a typical doctor you encounter. I have enough tools and medicines available to practise like a normal doctor, and, we also offer healing of the mind."

At this stage Nisha locked in her decision to leave the practice. It was now a matter of how to make her exit, not when. Her commitment to herself to not participate with Healer Milar brought confidence back to her voice. "Um, my shoulder's hurt, not my head."

"Yes, it is." Milar paused. "How does your shoulder work?"

"It has muscles that help my arm move while it's connected to some part of my back and neck."

Again the healer paused. "Deeper than that. What runs your shoulder and your body in general?"

Nisha felt frustrated by this line of questioning. "Of course, yes, my brain. Yes, my brain does control my body's functions but I don't control my brain or my body's functions. I don't tell my heart to beat, I don't tell my lungs to breathe, they just do it. I think most people appreciate how powerful our minds are."

"What about the power of thought? What about the possibility of healing through the mind?"

Flippantly, Nisha waved the idea away with a cheeky smile. "Normal healing would do fine for me."

"Would it?" asked Milar. "You could have easily ended up at a normal doctor yet you found yourself here."

With no clever answer she decided it was her turn to take the conversation off-track. "Why didn't the publication out in your waiting area make much more of a deal about the bombing the other day?"

"Hmm, I guess it's because they want to spread a positive message such as our words about healing and harmony. We want people's minds to be open to new ideas, and the events of the other day, well, the bombing might hold people back from larger contemplations. When you are stressed do you have a great clarity of mind?" asked Healer Milar.

"Not really."

"There are no mysteries in what we are trying to do. People have been meditating for thousands of years. Longer than most religions have been around." Milar was reasoning with her in a way that seemed conclusive and undoubtable.

"That kind of makes sense to me."

"Of course it does. Do you not feel better when you have had a chance to relax for the day or when you are properly rested? Do you not feel a greater clarity of thought and purpose?"

"You're good you know," said Nisha wagging her finger at him.

Milar continued. "I think we are ready for more as a human race, ready for a further step. We can speed heal ourselves with the help of machines, high intensive energy machines for exciting our muscles."

"I know, I tried one in Charing Cross Hospital the other day; it didn't really help much."

"Just wait here, I want to show you something." The healer went to a back storage room and returned with what looked like a silver neck brace, only longer and thicker with various controls.

"What is that?" Nisha recoiled from Healer Milar.

"It's for the next stage of Universal Direction. For healing of your mind and your body, if you allow this."

"I do not allow this!" Nisha said emphatically, pushing both hands downwards to reinforce her refusal.

Healer Milar gave a reassuring glance. "That is fine. In any case you are not ready and possibly not truly in need. Ultimately, it could be a great help to you though. For now, I just wanted to show you the mechanics of the device."

"Is it safe? What does it do?"

"It's a brace we attach to the neck of patients. It helps them reach harmony. This is the mind-healing we speak of. Some are calling this Curate-mente."

Healer Milar turned to the man in the far corner. "Christopher, will you please come over for a minute?"

A voice came from the circle mat. "Yes, Healer Milar." Obediently the tall man pulled himself up and walked purposefully over to where Milar and Nisha were seated. He spoke in an almost vacant tone. "What can I do for you?"

"I just want you to place the healing device on your neck and then talk us through what it feels like."

"Yes, certainly." Christopher Harkness took the intricate-looking neck brace and returned to one of the oval mats set within the square area. Sitting cross-legged, he placed it firmly around his neck in such a manner that told Nisha he had done it before. The silver device covered all of his neck and the lower parts of the back of his head. It was difficult to tell with the naked eye but the brace appeared to move upwards and adjust to make a perfect fit under his head. Finally, a long flat cord was connected to the back part of the brace that wound its way back along the floor to a half-opened cupboard that Nisha could not see into.

Healer Milar reached to a point on the side of the silver brace and placed his finger on an indeterminable area to start the mind-healing device working. It commenced with a click and hum of power.

This was happening all too soon for Nisha. Just a few moments ago she was seeking some assistance for her shoulder. Now she was watching something she felt was private, intensely so. It seemed like it shouldn't be seen by her, some stranger off the street.

Milar stood back from the man with his hands on his hips as if admiring a table he had recently built. "This is his eighth time. We set the charge low so the penetration is under the control levels."

"What do you mean charge?" Nisha's discomfort levels were rising more and more.

"It's quite safe, although we do push a fair amount of non-conductive energy into certain parts of the mind."

"What? Um, no!" Nisha arose quickly and went straight to the door.

"Please wait." Nisha heard the words from behind her. Not from Healer Milar but from the pacified patient. She stood there with her back facing the men behind her. Christopher Harkness spoke again. "There is nothing to fear here. I am perfectly fine. More than that, I am completely better. I've never felt more in tune with the world around me than now. I sense your concern. You're worried that this moment could escalate into something more impactful, whether now or in the future. It won't. I can promise you that."

Nisha turned around to face the man and did not say a word.

"I've had a difficult life. It has been hard to remove it from myself and my memories. You can look me up if you like. I've had many appearances before justice for my crimes of violence and theft. I did these, in hindsight, without reason for I did not need food or money. I realise the impact of my actions on others. I feel them. Stay and watch, it's perfectly safe."

Against her better judgement Nisha walked over to her stool and sat down again, shifting it backwards a few steps from the mats in the middle of the room.

Healer Milar continued with his demonstration. "It is a two-stage process: first, the brace accesses the amygdala and in the simplest terms starts a form of management. This is essential to the next stage as we need the amygdala to make sure there is no fear or anxiety in the patient. We basically help them assess what is to come next as non-threatening."

"That sounds like you are sedating them." Nisha was doing her best to look unconcerned and to stop herself from running out of the room again.

"Then the brace starts a process of opening up the pineal gland."

"Okay." The event in front of her was clearly no longer normal. The situation was vastly different to any moment she had been in before.

Milar gestured at parts of a brain in mid-air. "Through this gland, this individual gland, we can bring the two sides of the brain together, to bring harmony to themselves. Through this we can show the patient they are more."

"I'm lost. What do you mean?"

"The experience Chris is about to have is going to be one of intense light. Of a lifting off in his body. He will feel weightlessness; he will feel warmth. Some people describe the feeling as harmony and some people describe it as love, pure unconditional love on a level that they cannot fathom, let alone measure."

"This sounds awfully prehistoric, like something before the Quadrant Period."

"The process from the brace is a closed loop; we are motivating the amygdala to speed up the retention of this memory from short-term to a longer-term."

"That all sounds like manipulation to me."

Healer Milar looked unchallenged. "What is the energy healing machine in Charing Cross if not energy manipulating bodies? What is massage? What are medicines like sedatives or drugs for the brain's neurotransmitters?"

"This seems different. It's playing with parts of the brain I've never heard of being played with."

"Maybe so. We are progressing forwards in humanity. After all, the past is the past." Healer Milar moved his attention to his patient. "About now, Chris should start to feel a warm and ever-so-slight vibration in his head." In response to this description, Christopher Harkness nodded. "Then he will experience an increase in the vibration and hopefully an exponential jump in his body and awareness. To such an extent that it will take his mind away from the normal sounds and patterns it is used to. For a period of time, Chris will feel on a different plain of existence to us. He will be lost in light."

Nisha stared at the man sitting cross-legged on the oval mat. If not for a few small twitches and movements in his face and body, he was perfectly still. Again she felt the guilt of watching this private moment. Possibly, she should turn away while it was happening. As if the man was naked in front of her. Then, with a slow curve of his back, Christopher Harkness fell backwards and his legs were drawn down. No more did his face twitch or his body move. He was perfectly still.

"He will be like this for only a short time, a few minutes at best. This is all it takes to feel better within oneself. This is in front of you, the second stage of Universal Direction, Curate-mente. It is my belief that if we follow this path we can all achieve the final stage of liberty of consciousness."

Nisha wanted to relax and hide her expressions from the healer. However, she could not contain the look of puzzlement on her face. "I understood the 'liberty of consciousness' to be a slogan or election campaign, like a call to open your eyes and choose something different."

"No, it is real. In truth it is something truly difficult to obtain. Through meditation and healing of the mind we can get there. We can all liberate ourselves from this delusion we consider reality."

Nisha did not know what to say. Part of her wanted to laugh, part of her was curious. The Universe was big and boundless; why could it not include such an idea? To her, however, it felt far removed, and watching Christopher Harkness lying deadly still on the mat convinced her that she was not ready for Universal Direction. "I'm sorry, I've had enough. It's time for me to go anyway."

"Go with light then, Nisha." Healer Milar smiled and acknowledged her exit, almost finally allowing it.

Nisha felt relief from the release of this situation, so much so that she could even endure the last earnest statement from the healer. "Yes, okay. I'll do that."

In the sunlight, walking back down Queenstown Road, Nisha realised she needed to clear her head and understand what she had just seen. Officers from the Agency would still be there following her. After witnessing such an intense moment she felt okay about the scrutiny. The Agency would follow her and watch over her as she made her way around south London and over to Wandsworth Town to see Jasiah.

By the time she arrived at number 28 Marcus Street it was late afternoon and the heat of the day was dissipating in the twilight. Nisha knocked on the door and received no answer. She was concerned. Jasiah very rarely left home and of late she had not heard from him. A long time ago he had provided her with an access code for emergencies and this situation seemed to qualify as such. All looked normal inside except for a musty smell that told her windows and doors had not been opened recently. She moved past the bedroom and kitchen into his small lounge area and the large view screen on the wall activated in recognition of her presence. It was set to do so after many nights watching vids together and sharing music.

She gave it a direction. "Open news stream." Instantly a steady pace of updates came up from sources that Nisha was not familiar with. These updates differed from Jasiah's normal feed of news from home and nature channels from the outer colonies. Rather, the focus was on Earth, the bombing at St Paul's and Universal Direction. Nisha looked around the lounge-room while the news stream periodically updated. "Where are you, Jasiah?" she spoke aloud. "Open calendar." Instantly, Jasiah's location was revealed. It was in south London, Mordon. There was an event titled "Montaxe-Deus London 2". It was starting that evening.

Nisha entered the name into her calendar; the details presented to her were confusing.

Montaxe-Deus – Assembly Two

Brother and Sisters come together to battle the disassembly of our way of life
All brothers and sisters, from all gods and goddesses
Montaxe-Deus will represent you and act for you

Morden Hall, Morden Hall Park,
Morden
Wednesday, 22 October, 6pm
Bio-T screening on site

Nisha left Jasiah's house with the details for the Montaxe-Deus meeting copied into her communicator. Straightaway she spotted a silent follower from the Agency on the far side of the road. Nisha could not be certain, but it looked like her original interviewer from the Agency, Sean Watson. She looked down at the communicator on her wrist that held the recently downloaded details for Jasiah's location. Sean Watson mirrored her action and read his communicator. He looked back at her and shook his head. Nisha started to suspect Montaxe-Deus wasn't something the Agency, nor Hames, would approve of.

Chapter 15

Positive Passing

HAMES ENTERED THE QUADRANGLE TO find the lights dimmed. From what he could tell, all of the crew had gathered around the edges of the large oval shape that filled the capacious room. The elliptical window above them correlated to the form below, showering the shining stars down on them. A deep ring sounded, the sonance of a crystal metal bowl being struck. Then another, a bowl of lighter tone. Both sounds persisted and entered his chest, vibrating in his body. For the briefest of moments, he forgot why they were all there. He forgot what had happened to Xavi and Chen Chung in the repair engine room.

Of course, he would be last to arrive at such a gathering. It was unnatural for Hames to be drawn to this warm and airless space. In truth, he felt repulsion. He knew what was coming. This was to be no funeral and it would be a surprise to see any tears flow. Hames had been to other Universal Direction wakes and knew what was about to unfold.

Emmalyne turned to see him and he was jolted by her appearance. Her long blonde locks were unkempt, framing a set of strained hazel eyes. It had only been two days since what he reluctantly agreed to regard as an accident, and she was still vividly in a state of shock.

Hames took a spot on the mat across from her. In physical solidarity with his young colleagues, he felt mute with only devastation in his heart. By contrast, the faces around the room were content and full. Most of the crew were smiling and conversing as if this was just another meeting of the Mission and Operations teams. He was comforted to see Emmalyne at least was feeling strong emotions just as he was. Without looking or giving notice, she arose and moved to sit behind him. Hames felt her two arms crossed and resting on his back with her hair falling around her face and his shoulders. This closeness was soothing and reminded him of piggy-back rides he used to give to Nisha's youngest sister on warm Mermain days at the beach.

The chimes of the metal bowls continued to ring, unrelenting in their penetration of his skin. The slow tempo of the ringing seemed to allow amiable discourse to continue in a free-flowing nature. Hames could hear their names now, Xavi and Chung. He also heard the fourth Universal Direction convention more than once: "you are the everything". Then, from a group sitting further away, he recognised the derivative version of this maxim in the conversational context: "they are the everything". It was a comforting thought. So comforting. He wondered, as he had done at other times, why he could not have this comfort. The concept that Xavi and Chung were not gone, just not present, was a sound way of coping with death. It just wasn't within him to accommodate this view.

Yolanda stood up and looked as untroubled as usual. "You gotta look at their passing as just part of the ride. Ups and downs." The chief engineer moved her arm in a wave-like motion as she stepped to a panel in the centre of the far wall. She pressed her hand against a small illuminated panel. Projected into the centre of the room came a star that was visible despite the light from the Universe above. It was Sol, Earth's sun. The image began to emit warmth and light. This felt entirely nourishing to Hames. For a moment, guilt came over him as in most instances he deemed the Universal Direction praise of the sun as religious in its deference. He knew, though, that he wasn't here for himself. Xavi and Chung were completely dedicated to Universal Direction and his presence would have made his friends happy. There was no denying it; sitting in close proximity to his colleagues, under the warmth of the sun, made him feel better.

Hames' guilt at the peace offered by this close union soon faded to be replaced by anger. Two days ago everything was fine. Now, two of his friends were dead and the *Medens* was broken. Non-functional, just like the space it was meant to repair. Between Leena moving the *Medens* away from the debris and Emmalyne bringing the repair engine back in line with the ship's systems, they had averted total destruction. The key question was, *why had the repair engine failed so badly?* Why had the interface block failed to manage the situation? Xavi and Chung's last moments were spent applying all manner of effort and tools to avoid the disaster, to no avail, just their regretful passing. With a lack of information or a clear answer as to why it occurred, the event remained unclassified. Hames was so far unable to tell whether it had been purposeful or simply a failure of luck. Overall, however, there was a pattern emerging. First the light alignment drive accident and now this.

Around him the celebration of Xavi and Chung's lives continued.

"What will be will be. Yes, Xavi would want it this way," Temett said in brazen happiness. His shoulders wide and content, matched only by his relaxed smile.

Even Robero Paldini, clearly still ill, was mustering his energy to acknowledge Xavi and Chung. "The past is the past," he said with a weak smile.

Temett continued to nod. "They chose this, they chose this path."

"That's right." Yolanda was also in high spirits. "We all chose our destiny. We may not know it but what will be, will be."

Francois looked like he was just as content as the others. "We are all together now, here in this place. We are part of everything. Xavi and Chung are part of us."

Hames felt Emmalyne shift her head back and forth on her crossed arms as she lay on his back. Francois puzzled him. Both were older in years and from a different time. However, Hames had stayed in that previous age, a time before Universal Direction, while Francois had grown into the post-Cataclysm world. The French-Cernunnosian had fully moved with the new way of societal harmony.

Throughout the wake, Olisa sat silent and beaming. Hames felt his mouth curl in on itself for a moment. It was like this was a party especially for her. He felt she was enjoying it far too much. Sitting next to her, Dao Mai, the plainly spoken astronaut, nodded in perfect alignment with his friends. Others continued to tell stories of their time with Xavi and Chung, all without a hint of sadness. Yolanda continued, exuberant in her recollection, although the COO sitting next to her was quiet except for starting the proceedings with a few words. Leena moved along with the beat of the group, neither effusive nor obviously sad.

Hames recognised that death should be a time for celebration. He assumed that most people hoped they would be remembered fondly after death but wanted their loved ones to move on in life. Hames had not been successful at the moving-on part. In fact, he had done the opposite and dwelled on the loss of those around him. He considered this was acceptable as he had mitigating circumstances. Nisha might still be alive somewhere and letting go equated to letting her down.

Finally, the COO walked over to the panel on the far wall, her steps strong and body upright; even in tragedy her whole demeanour never betrayed the truth that she was the leader on the *Medens*. She faced the crew with her deep-olive–green hand, pausing before activating the panel beside her.

"You're all welcome to take more time throughout the coming days to reflect and consider Xavi and Chung. Unfortunately, I have a ship to run and I need my crew to grab some food and move to it." Melini Guerra paused to allow this news to sink in. "Our ship is essentially unusable for what it's meant to do." The COO slowly switched off the projection of the sun, allowing just the chimes ringing in the quadrangle to continue. Hames sat there with Emmalyne resting on his back, uncomfortable with the world around him in the absence of the light of the sun, and the departure of Xavi and Chen Chung.

After some coercing, Hames took Emmalyne towards the mess hall, hoping food and the companionship of those not on duty would help her process what had happened. Regrettably, only Olisa and Leena were present. Both were involved in a discussion, which with its open gestures of hand and face indicated satisfied regard for the proceedings just held. Leaving Emmalyne with these strong proponents of universal harmony, Hames attended the food dispenser wall to withdraw items of comfort for his young friend. Rejoining them, he caught the mid-point of their conversation as Emmalyne explained her thoughts.

"I'm really confused. I want to be happy for him. But he wasn't happy, not at all. Xavi said so to us before he passed he…"

Olisa reached out her arm, its age revealed by her grey rolled-up sleeves. She took Emmalyne's hand in her own. "I have taken the liberty of booking some time for you to talk about it, young miss." Olisa stared at Emmalyne earnestly. "Xavi has gone from this place, this plateau of matter and singular focus. He's gone from nature to the Universe in a beautiful liberation. It is truly amazing. I've never heard of someone being taken in such a manner, just incredible!"

"That does sound better than what I thought happened to him." Emmalyne's shoulders lifted somewhat at Olisa's explanation, although she still looked disorientated by this view.

Hames was exasperated. He could take no more. "What are you talking about? Listen to yourselves. Xavi is dead. Doesn't this upset you, make you angry?"

Olisa did not let go of Emmalyne as she turned on him. "Chief Hames, I think you need to take some time for yourself. We all know how you feel about our practices, now is not the time!"

Leena implored him as well, in a gentler fashion, speaking quietly and almost pleading. "Chief, not now."

Tears welled in Hames' eyes, abated only through one last remnant of control. "You make it all sound like this was destined to happen, that Xavi and Chung chose this! I can tell you they didn't want this at all!" He felt truth for the first time that day. In his contrary position he began to grieve and accept the passing of his friends.

At the table for four, Olisa closed her eyes to him as if to show her indifference to his views. Leena, however, was not in the mood to ignore him. "Xavi was my friend and he's gone. I'm not happy he's away from me. I can't hide that. When this happens, though, you can choose to remember him in pain or in joy." She offered two hands to Hames to represent the alternative choices.

Restraint abandoned the chief mission officer. "There was no joy in the repair engine room! Xavi and Chung are gone forever. They're not some element of a greater being returned to its source. Our two friends who walked and talked with us are no longer able to do that."

Leena tried to calm him. "I'm with you on part of it, Chief. The Savai view is that we retain ourselves after we pass. We don't merge back into some conglomerated mind mess." Olisa raised an eyebrow at her young colleague. "Yes, this accident took my friend Xavi away." Leena turned to look at Emmalyne. "I just choose to believe that wherever next he goes as Xavi and I'm sure it's somewhere beautiful."

Emmalyne then put her head to her hands and let her tears fall. "I just feel so so bad; if only we had worked quicker. If only I had worked through things quicker and more efficiently we could have fixed the problem before it took them!" Emmalyne was hyperventilating. "Before it took Xavi!"

Leena put her arm around her friend while Hames struggled with his awkwardness. This difficulty of emotion had persisted despite his connection with Emmalyne. A connection reinforced through sharing the passing of someone, a person who had reached into death by way of returning to energy.

Olisa began to talk more to Emmalyne, comforting her in a way that Hames realised was beyond him. "My young girl, we look to the Akasha. The soul space of where they are now. We are here together with them in the Akasha. They are not gone, merely liberated." Emmalyne sat quietly in response.

Leena tilted her short-cropped head to one side, indicating the door from the mess hall. Hames stood up and accepted her offer. As the exit point closed behind them, Hames shot a shadowed glance at Emmalyne being consoled by Olisa.

The small first pilot's distinctive nose pointed them down the corridor. "Chief, I know I'm a lot younger than you, well, a massive amount younger than you..."

"Thanks." Hames appreciated the cheekiness and the walk they were on.

"Have you noticed, though, that it's only you who feels the way you do? You're suffering, and perhaps confusing Emmalyne a whole lot."

Hames didn't know what to say. Normally he disregarded Leena's manner as somewhere between argumentative and misplaced righteousness.

"Don't you think it would be a better use of your energy to look at the reasons for what happened rather than to rail against everyone else's feelings? Something has gone wrong on the ship. Are you going to figure this out, do something positive for the people around you or will you just focus on yourself and what you want?"

Still he said nothing. Below their feet, the energy conduit for the ship enclosed in the gangway lay dormant. No longer did it pulse with energy and hum its binaural tones to remind them of life within the *Medens*. For the first time in years Hames realised how solitary he must look to others. All he could do was look to the practical issues ahead of him.

"We're missing something with the repair engine. It's not a part failing. There's another variable in there we can't see or are too blind to see."

Leena walked on beside him, stretching her small step to match his. "Maybe someone on the ship caused this; maybe it was sabotage."

Hames struggled with this idea. "Everyone is accounted for in the log, all the system security, the bio settings, the access points. There's no-one on board who could have done this or wanted to. Outwardly, it looks like a technical failure."

They arrived at the home-pods. "Remember what I said a few weeks back, Chief, when the Ugor Eels were here?"

"No, I'm not sure I remember."

"Don't forget you're a part of this crew. If you make decisions for other people or without other people you could end up pushing them away."

All Hames could do was nod and make his way to his pod to rest and refocus his mind on other matters.

That evening, Hames found himself in the far corner of the storage room. Alongside him Melini Guerra, Yolanda, Siggy and Olisa stood in silence. Theirs was a task assigned to them under the Organisation's Ship Management Procedure. The five of them were to oversee the disposal of Chen Chung's body.

"I can't believe we have to do this." The portly and concerned Siggy Conson shook his head twice. "I've read about it and watched the system work a few times, not for real though." An intense sweat had broken out on his forehead. "I can't believe it."

The group of five stood over the manual disposal unit that was large enough to hold a human body, much like an old-style coffin in size.

Yolanda was far less exuberant now she was extricated from the Universal Direction wake for Xavi and Chung. "I can't believe that we don't have Xavi's body to deal with."

"Hmm, yes." The COO paused and looked across the rectangular-shaped box to Hames. "So, Chief, any sign or trace of the body? How do we explain it in the ship's log?"

"Like I said in my first report. It was all a bit of a blur. Alarms and lights blaring. On one hand it looked like Xavi disintegrated when he touched the PCD matter. If I'm truly honest, it looked more like he was consumed by it."

"He is now everything." Olisa spoke without looking at the others, staring down at the shape in front of her. Hames had an overwhelming feeling to curse at her, but checked himself.

"Okay. Time to say goodbye before I push this button here and send Chung's body away," said the COO.

Olisa placed her hand on the disposal unit. "We are here together, my friend, always."

Yolanda's send-off to Chung was also a repetition of Universal Direction convention. "What will be, will be."

Hames placed a hand on the top of the disposal unit.

With a push of a blue button, Melini Guerra released Chen Chung's body to space. It shot away at a great speed and vanished from their view in seconds.

The chief mission officer remained stuck to his spot as the others walked slowly back through the exit point that opened at their presence. None of the four departing members looked back to Hames and the disposal unit, his presence forgotten. The exit point closed, leaving him small in the quiet of the ship. There were moments on the *Medens* when it was very hard to believe Leena's directive that he should feel part of this crew.

Chapter 16

Moon of the Kraken

OVER THE DAYS THAT FOLLOWED, Hames wandered around the *Medens* performing obscure tasks in rarely visited locations. The overcome repair engine and the *Medens'* attempt to violently tear itself apart meant the mapping of structural damage throughout the ship was an urgent priority. During the incident, Hames had heard multiple cracks and shattering of metal. So far, the damage he'd assessed was thankfully minor and manageable. He finished examining the storage area and began his way to report to the COO.

On the bridge, a lack of activity on the screens and panels told him the seat of command was more subdued than normal. The *Medens* was in containment mode. With the breakdown of the repair engine, the ship's major energy source, all recreational services were offline. Food units were down to providing the most efficient constructs, and in small quantities too. Hames did not even want to think about having to consume the nutritionally rich cardboard-like rations sequestered in the darkest parts of the storage area.

At the helm, Leena was overseen by a tired-looking Melini Guerra, both of them flicking through manuals and checklists that served as guides for problem solving. Leena sat in the pilot's seat and the small white paddle reached out from behind her to rest on the gold spot on her neck. She spoke to the chief engineer, who was operating the derelict light alignment drive. "You right to go again, Yolanda?"

A voice returned to the command centre. "Yep, the alignment drive is ready whenever you are; just send it the order."

"Do it," commanded the COO.

Leena turned her head to Melini, as much as her physical linkage with the *Medens* would allow. "Can I count down? You know I like the countdown." Leena's distant cousin and commander raised one eyebrow wearily as she sank back into her seat.

"Okay, no countdown then." Leena pushed her hands through the control panel to engage the light alignment drive. The *Medens* performed its customary jump towards matching an immeasurable point near the speed of light. However, after a few seconds it retreated from this effort back to its normal cruising speed.

The COO banged both her fists hard on her chair. "We're getting nowhere with this!"

Leena withdrew her hands from the controls as the connection paddle detached from her neck. She breathed with resignation. "Well, we're actually getting somewhere. We're not too far from New Galapagos with all these mis-jumps."

The dark rings around the COO's brown eyes were all the more emphasised by her deep Mermain skin, which looked more akin to her age rather than a youthful mature woman. "I can't tell if you're being positive or not, Leena. I am that tired I have no idea."

"I'm a good measuring point for these things," said Leena. "Perhaps it's time you took a break."

The COO shook her hands by her sides before resting them on her hips. She smiled and Hames thought it must have been for the first time in days. "Right, I'll get some rest then. Before I do, Chief, tell me the status of the repair engine."

Hames started to speak, his voice catching in his throat. Already tired and drained before the accident, the health of his body was fading. He tried again. "It's not good. Seriously. It's been beaten about by the overflow of particles and stress. Lots of parts need replacing and it'll take most of our spares. Also, the main conductor plates need reconditioning to match the rest of the repair engine components. That'll take some time to do. We can't put it all back together again until each section is ready to process particles at the same frequency. Even then, we need to access some areas from the outside, so the *Medens* needs to be stationary."

Melini Guerra pulled one of her cheeks downwards in despair. "Sounds like a job to do at New Galapagos. The cause?"

"I still don't know. I can't see any purposeful damage. Something triggered havoc with the interface block. Its functions were completely nullified." Hames held his words for a moment as he rubbed the lower part of his grey-flecked beard. What he was about to suggest sounded more like the mysticism of Universal Direction and not exactly empirically sound science. "If we also consider the accident in the light alignment turret room, something looks wrong. One accident is a coincidence, but two? Well, that's another matter. Something is definitely challenging the mission."

"Maybe," the COO pondered. "The *Medens* is a volatile ship. If someone wanted us destroyed they could've done that many times over already."

Leena sat back down in her pilot's seat and grabbed one knee to her chest. She gave some credence to Hames. "It's a reasonable proposition in the absence of any more data. The only sensible response, though, would be to stop the mission. We shouldn't risk any more lives if we really suspect we're being threatened."

A wave of anxiety washed over Hames. This truly was his last chance to reach out to the Universe to find Nisha. The *Medens* turning back would be a personal disaster. Leena was probably right and he had helped her reach this conclusion. He needed to buy himself time to show there was no danger to those on board. The chief mission officer once more cleared his throat. "I think we really need to break it all down and examine what happened before we jump to conclusions. This all feels quite arbitrary to me. We'll know more once we get to New Galapagos."

Melini Guerra turned to Hames, looking strained. "We do seem too close to turn around don't we? Either a few weeks to the planetoid system or we turn around now, travelling back to Earth without any spares and no energy to create them. Chief, can it even be fixed, the repair engine?"

"I don't know. I need a second set of hands and Emmalyne doesn't really want to go back in there. Not after watching what happened to Chung and Xavi."

As she looked out to the black space ahead of them, Melini sounded despondent at the loss of her ship's capacity. "We need to get the repair engine back online as soon as the other parts are ready. Once we make it to New Galapagos we can start the work. Then we make a decision whether we push on repairing the tentacle or we head back home to Earth."

Leena gave a nod of agreement.

Hames gave his understanding with a quiet "yep" and a scrunch of his mouth. He realised having a repaired *Medens* back at its full purpose, fixing the damaged space and heading to Mermai, was just as important to the COO as it was to him. How else would she hear from her family? They both wanted to find out what happened on Mermai all those years ago.

The COO returned to her manuals and the exit point for the bridge shifted to allow Yolanda's arrival. Hames noted the normally refined chief engineer looked bedraggled and rigid in her body movements. Yolanda was under pressure for the first time in many years and was struggling to find an answer to their predicament.

"It's still stuttering isn't it?" Yolanda directed to Leena.

"Yeah. You saw it. Just a few seconds then we're drawn back." Leena made a motion with her hand as if to show the *Medens* slowing to a backwards crawl.

"The structural damage caused by the repair engine overflowing is causing the LAD to lose its alignment. It's going to keep dropping back to cruising speed, that's my call. Without the extra power from the repair engine I can't get it to hold light alignment for more than a short period of time." Yolanda was extremely frustrated.

Leena turned back to her screen and raised her right arm with two fingers pointing upwards. "That's two forms of interstellar travel we've lost now."

Hames thought of the half-wrecked repair engine. "I've never seen something like this happen before. I've only ever seen unconstrained particle conversion debris up close once before. It wasn't long after the Cataclysm. The Organisation deemed it far too dangerous to ever get close to."

Leena moved her seat around. "Do you ever think we could use PCD travel again, Chief? For real? To get back to the colonies?"

Hames looked at the COO then back at Leena. "The theory from the fracture scientists is that the cracks are biggest and most prevalent around Earth. So that the further out we map and repair the tentacles, the more likely it is we'll find clean space in which to travel. It's wishful thinking, though, as everyone understands it would be fatal for a ship to travel using particle conversion tech. If a ship hits even the smallest of fractures it would be decimated, and no-one knows just how far and deep the damage has spread at the other end."

The COO grabbed their attention again. "Chief, we have another problem. If we get the repair engine fixed, who is going to maintain it and the interface block? We don't have that kind of personnel redundancy."

Hames was lost for an answer. "In a worst case scenario we would lose Chung or Xavi, not both at the same time."

"Right now it's just Dao, Emmalyne and you. It's not enough, and like you said, Emmalyne hasn't been helping out much lately. Spending more time with Olisa I see."

Hames ignored this part. "Well, I'm not sure what to do. Maybe Robero or Yolanda could help out."

"Hey, didn't you just hear me? I have my hands full!" exclaimed Yolanda.

The COO put her long brown hair back behind her green Mermain ears and casually comanded Hames: "Chief, you have to find someone on New Galapagos."

The chief mission officer was genuinely perplexed with this request. "I don't see how that can happen. Is it an option?"

"Well, it's one of the options. I want the *Medens* to have a chance to make it a few light years past New Galapagos before we eventually decide whether we can make it to Mermai or not. Hopefully by then they have solved this problem, and fixed the tentacles. Or maybe, just maybe, at that point we can communicate with Mermai."

Hames could see the hidden desperation in her eyes. The COO had her whole family to look for; she wondered what had happened in their lives. She could be a grandma by now and she wouldn't even know it. "Okay, I understand."

"Chief, offer them a five-year contract; do whatever to make it worth their while. They can own one of the New Galapagos moons for all I care. Just get me a new crew member. And a good one!"

More weeks past and the crew finally settled into a new pattern as they made on and off again jumps to light speed. At first this constant shifting was only mildly tortuous. Ultimately, though, the repetitive jolting from the transition became worse and worse with each sporadic jump. This paradigm of existence made the entire crew feel off balance.

Added to this the power levels on the *Medens* had started to diminish with the loss of the repair engine. The recreation room and the Williams Sphere were off-limits and only mission critical parts could be reprinted and built. It had become clear that it was imperative the *Medens* reached New Galapagos.

A short distance away from their destination, the light alignment drive failed entirely. No longer did it provide short sprints of travel. Fortunately, this was not the calamity that it could have been for the *Medens*, being only five days from New Galapagos at cruising speed. The day after the failure of the LAD, Second Ship's Pilot Temett and the light alignment crew of Yolanda and Robero joined Hames and Emmalyne in front of one of their large mapping screens to examine New Galapagos and its four moons from afar. Hames was thankful that in the last week Emmalyne had returned to duty, spending more time in the repair control room preparing an outage plan to fix the repair engine. His young colleague was trying to focus back on work after the tragedy of Xavi and Chung.

"You're right, Yolanda. They look like big rocks sitting in the middle of nowhere; the second one is almost elliptical," said Emmalyne.

Temett stood behind them all. His strong chin pointing towards the cluster of four large rocks around the planetoid of New Galapagos. "From the scanners it looks about five days until we reach the fourth rock or fourth moon." He looked at Yolanda for confirmation."

"Ask Hames, he went to that moon." Yolanda directed her elbow at Hames while she studied the large space map intently.

Hames stuttered, "Oh, I can never remember what they were classified as. They're not moons but people call them that. So let's go with the fourth moon."

Emmalyne was able to dissect the arrangement and the path of the *Medens*. "The fourth one sits high compared to the others; it looks like we'll pass right under it as we cruise into New Galapagos."

Yolanda nodded, a rosiness finally returning to her complexion now the light alignment drive was permanently offline. "That seems about right, and it's the normal path into the system when coming from Earth."

Temett continued. "We'll have to switch off cruise mode a while before then and coast into New G. When we're drifting we'll sail slowly below it, like a boat under a bridge."

With her American inflection, the chief engineer agreed. "I see what you're saying; we'll get a good look at it from underneath. Okay, organise it, Temett."

Four days later a message went out regarding the fourth moon. That evening most of the crew made its way to the quadrangle where they all laid down, looking upwards through the large oval portal that showered them in stars. Hames felt as if he was about to participate in a Universal Direction session. Olisa's absence at this gathering told him there was nothing to worry about on that account. He assumed that her nonappearance was, in part, due to the usage of the quadrangle for a purpose other than Uni-D. However, this was only a guess. Eight members of the crew lay in the dark beneath the window as the *Medens* cruised onwards towards the planetoid of New Galapagos, waiting to pass under its fourth moon. Hames found himself next to Emmalyne, who was next to Leena.

Their peace was broken when Yolanda gave some context. "When I was at New Galapagos I heard the old stories about this fourth moon. Apparently every few years when rare solar winds hit it, the moon lit up like a greenish-blue ball and almost pulsated. Parents apparently tell their children there is a Kraken in there, a giant sea squid–type creature that will come down and eat them if the children don't do their chores," she drawled.

Hames thought back to his own time on the moon. It wasn't a pleasant experience and one he did not care to think about, let alone explain. Besides, there was a feeling of serenity in the crew as they coasted along, lying down in the night. Harmony without the chimes and conventions of Universal Direction.

"Look, here it comes!" shouted Emmalyne.

"Nice one, Emmalyne," grumbled Leena, screwing her finger in her ear.

Temett echoed most of their assesments. "It's sharply scarred, lots of cliffs and valleys. I expected it to be more crater-like, more like our moon."

"They should call this moon Verona," suggested Emmalyne.

"Why's that, young Emmalyne?" asked Francois.

Robero Paldini understood. Despite his ongoing weariness, there was an Italian ingenuity about him and this showed in his tone. "Ah yes, Emmalyne, very good. I get it."

"Help me out, sick boy." Yolanda nudged the engineer next to her.

Robero complied. "The jagged nature of this moon, and its very tall cliffs. Much like those found on Miranda, a moon of Uranus. Verona Rupes are the largest cliffs in our home solar system, over five kilometres high."

"Doesn't this moon have a name already?" asked Temett.

Dao, who was also lying with them, provided his useful factual insight. "Version AR2582 of the Organisation Star Charts lists this body as NG-004 or commonly known as Four to its people."

"What a plain description." Leena evaluated the local naming effort. "Emmalyne's suggestion is much better."

"I'm with Leena on this one," Yolanda agreed. "Verona Moon it is."

With the christening of the moon gliding into their view, they continued to stare contentedly. Hames felt the peace and quiet of this image creep into him until he saw a small green fleck of light flicker for a fleeting moment from the surface. He felt a tiny moment of concern beat back the serenity he had just felt. However, no-one else seemed to notice this light; perhaps it was in his mind. The others seemed calm and it did feel a moment to try to let peace in.

For Hames, though, both apprehension and excitement were growing inside him now he was returning to New Galapagos.

Emmalyne appeared to pick up on this unspoken discomfort. She turned her head on the mat to face him and whispered, "You okay, Chief? You looked worried for a moment there."

"I didn't have a successful time when last on New Galapagos. I went there after the Cataclysm for answers and found none."

"Maybe you'll do better this time."

"Maybe." In the years that passed Hames had learnt more and more about the broken space that had cut the Earth off from its seven colonies. Now he was returning with new calculations and new approaches. Once he had access to the biotower in the main citadel he would be able to scan the space around him. He could review the damaged particles for evidence of the *Python*, Nisha's ship. He felt trepidation and excitement that maybe more clues as to what happened to Nisha would finally be afforded him. He would also need to make sure the *Medens* would be repaired and sent back on its path towards Mermai.

The next morning Hames entered the mess hall to be greeted by Francois. "We're here, Chief, we're just entering orbit for New Galapagos."

"Okay. Have you seen Emmalyne?"

"No, I believe she is with Olisa."

Francois was correct; shortly afterwards Emmalyne and Olisa both entered the mess hall. Lately his working assumption was that the ship's healer was preparing Emmalyne for her first Curate-mente session. He watched this from afar as her trauma of watching Xavi consumed by PCD matter had persisted. He sat down uncomfortably with the new entrants to the room.

At a nearby table the COO's brown hair hung forward as she searched a screen in front of her for the data she was about to present. Emmalyne, in her plain-speaking fashion raised a question. "Why do you keep your hair long? It always seems to be bothering you." Emmalyne had not abandoned her curiosity about Melini Guerra's past and a return to her cheeky quest for information signalled a return to health.

Melini Guerra looked up at Emmalyne. "I like it that way. For me, my long hair is a reminder. Back home, back with my family, my daughter Miena loved my long hair. One day I cut it, really short. I loved it, it was a moment of freedom that haircut. I'd kept it long since I could remember. Well, my mother had kept it long and I had continued that. Yet, that day I came home with the shortest of short hair...oh, Miena's face, oh the look of horror on it! She was only four but she stomped off, just for a few minutes, then she stomped back and came up and shouted, 'I don't like it! You're not my mother!'" The COO was laughing at this as were the people around her. "After that I got it cut short just one more time to show her that a little girl couldn't behave like that. My husband and son were not so keen on it either, so eventually I let it all grow back."

"Was she happy?" asked Emmalyne.

"She was. One day about a year later, Miena came up to me and said, 'Mam, I love you, no matter what hair you have. But I do love your long hair and I want mine to be long like yours, always'. After that I couldn't change it."

Yolanda put her hand on the COO's shoulder, who opened her brown eyes wide and clapped her hands together to signal the move to a more formal conversation. "Right, Leena will be here in about five minutes after she has finished her checks for keeping us in orbit. She's going to fly the shuttle down for us; it's a small one and given you don't get shore leave ever I'm letting most of you come down. So it's going to have to be in two trips. The *Medens* will be fine up here with a small crew, a shuttle operator and someone from the Mission Team to work on repairs." The COO motioned to Siggy Jenson sitting next to her. "The ship's manager is sending you the shift schedule. No swapping and that's final."

"How long are we staying here?" asked Robero. He had improved over the last few weeks, although he was still dreadfully pale and somewhat sub-human. "It's going to be the last stop for a while."

"We'll have eight days, eight nights." The COO pronounced this quickly and her statement was hit with a round of groans and shaking of heads. It was not a popular decision, but the look on the face of the ship's commander told them all it was a firm one. "You all signed contracts, very generous contracts with the Organisation. You and your families are being well taken care of: 'generational benevolence' I've heard some of you say along this trip. Look, we'll only need a couple of days to restock and the fixes to the repair engine should progress on time." A sideways glance to Hames was returned with a confirming gesture.

"COO, please. Please can you reconsider? We get so little shore leave," begged Robero.

The COO moved forward with her briefing. "Siggy's been liaising with the Administrators on the planet below and accommodation is all arranged for us in the central area of the citadel, the biotower. I know you've all been talking about this visit, finding out about New Galapagos. For clarity, I'm not your tour guide so I won't repeat any of that stuff. Just be mindful of the Eagleton Accord. This is their place, their rules; it's not Earth and it will be different. No-one is to move about the outer parts of the citadel by themselves; you must have a person accompanying you at all times."

"Surely it's not dangerous? That's a bit overboard!" scoffed Temett.

"At least two people at all times!" the COO shot back. "I don't know if it's dangerous or not, and you don't either. So let's be smart about this. Hell, there could even be language differences, New Galapagos is the melting pot of the colonies, don't forget that. It won't be a single-culture location."

The COO continued. "Okay, I'm leading the first group of officers down to the surface. All of you on Siggy's list, you've got ten minutes to get your gear. Then we meet Leena at the shuttle departure point."

Siggy led them through the winding grey corridors of *Medens* to the storage section, leading inside to a wall of hooks and metallic cupboards. "Okay, New Galapagos is warm and humid so you don't want to be wearing your normal greys and blues down there. There's also a lot of bugs and things. Sure, the biotower emitter does most of the work to keep the critters out but this short-sleeved jump suit is lightweight, aerated and has a pulse-coin in the left hip section to drive away anything else."

"Siggy, they're all white – they look ridiculous!" Emmalyne exclaimed.

"I wanted you all to keep cool." Disappointment shadowed his round face and his cheeks drooped.

"The sun-ray won't make us that hot! It's mainly for light, you dolt!" Leena sidled up to Emmalyne to join the debate.

"Well, I didn't know that. Does it matter?"

"It does actually. We'll look like these pretentious space explorers come here pure and good. I don't feel pure and good right now." Emmalyne's fist was scrunched up, holding the lightweight suit and gesturing at Siggy.

Siggy was retreating within himself, sweating and attempting to navigate the angry crew members before him. "I put the trim of your group on the suits, blue for Mission, grey for Ops." His weak offering did nothing to placate their disgruntlement.

"Don't worry, man," Leena offered in defeat, "don't worry."

Hames took his new suit as he left the storage area. "Thanks, Siggy, you made my day."

Chapter 17

Descent to Galapagos

THE FEELING OF EXCITEMENT WAS palpable amongst the crew members soon to leave for New Galapagos. As they gathered together in the Shuttle Departure Point, Hames could see a new energy in their faces; they looked more alive than they had for weeks. There was a greater sense of happiness, perhaps more so than that provided by the emotional support of Universal Direction. This was what Hames perceived. Then again, it could be wishful thinking.

The Shuttle Departure Point sat on the farthest right-hand side of the *Medens*, alongside the repair control room. It appeared to be a normal room with long benches down either side, funnelling towards a pressure locked door. The COO, Yolanda, Emmalyne, and Olisa were seated in a row – the first four for this trip. On the other side were Hames, Robero, Francois and Siggy Conson.

"Right, my group first and Hames' second," announced Melini. "After Leena takes us all down this will leave Temett on first shift to stay and watch the bridge while Dao stays behind to work on the repair engine.

"Okay," called out Leena, clapping her hands together. "It's going to be a bumpy journey through the New Galapagos atmosphere."

Francois was curious about their destination. "How does New Galapagos have an atmosphere so dense so far out from its sun?"

Dao, who was there to see them off, replied, "There is enough gaseous compounds and water to ensure that with the correct environment, a light but pliable atmosphere is released. The addition of tools such as the sun-ray and biotower make New Galapagos more than habitable."

Robero was solely concerned over his sleep pattern. "I just hope the sun-ray gets turned down enough to help me fall asleep."

Yolanda fixed the white uniform of her scruffy and drawn-looking engineering colleague, making him slightly more presentable. "All y'all need to know is getting to the surface isn't fun. It will be a shaky trip and we're supposed to be representing the Organisation in meeting the New Galapagos Administration. You're going to have to fix yourselves up when you get down there."

"Yolanda's right," the COO confirmed. "Remember, you're always on duty on the planetoid. Even if you don't have a shift or a task to complete you're representing the ship and the Organisation."

"Okay, the official stuff." Siggy Conson cleared his throat and stood in front of the pressure door. "The Mons Administration has told me after we pass through the initial landing process we'll head to a welcome ceremony."

The COO leant back against the side of the pressure door, her face ruefully questioning. "Siggy, you didn't mention this before. Ceremony sounds something more than a welcome party."

Siggy patted his round cheek a few times. "I know, but it can't be helped. The comms channel only stayed strong enough in the last day to have a proper conversation with Quartermaster Chirila from the Administration. They're very excited for us to be coming, let me tell you." The ship's manager's eyes focused on the rows of crew members, avoiding Melini's direct gaze.

"And, what does ceremony mean?" Melini got to the point.

"Well, New Galapagos is famous for its bugs isn't it? It has that impenetrable cover of dense vegetation on its outskirts; the heavy cover of ice and CO_2 has given rise to some pretty different creatures." Siggy waved his hand outwards to suggest caution to the others.

"Go on," the COO gestured with her green-tinged hand extended.

"Well, we have to take part in a local and historic custom to show we're welcome to the colony."

"Siggy, what have you signed us up for?" Leena laughed nervously while Emmalyne just shook her head at him.

"It's nothing, really nothing. There is this one bug that is prevalent and harmless called chelonoida. They're small flying creatures, no bigger than the palm of my hand. But they have an outer body appearance very similar to a tortoise, with a shell and paddles that are like wings." The ship's manager made swimming motions with his arms to express the chelonoida's capacity. "They're a mixture of rich colours – browns and greens. They really do look like flying tortoises."

"What do they have to do with the welcome ceremony?" pushed Leena.

Siggy scrunched his face, trying to think. "The Chelonoida are everywhere; it's hard to miss them. They can survive the cold of space or the heat of the sun-ray. So they were here when the first settlers arrived. They're the reason this place is called New Galapagos – the small flying tortoises reminded people of the protected park back home."

"You're leaving something out," said Leena.

"Well, apart from being pretty to look at they're fragile creatures to touch and..."

"And?"

"Quite nutritious as well."

"Ew!" came the response from more than one crew member of the *Medens*.

"The squashing of chelonoida is somewhat of a tradition," Siggy finally confessed.

The COO's demeanour was icy at this belated report from her ship's manager. "What does squashing this chelonoida have to do with us?"

"I don't know, I'm not sure...there'll be some there, something like that," Siggy stammered, a sweat breaking out on his forehead. "It's nothing. I'm sure it's nothing."

"Does the ceremony recognise Universal Direction?" enquired Olisa, who had approached Siggy's side in an almost stealth-like manner. Siggy was relieved by this alternate line of questioning.

"Yes, yes, it does. The colony received lots of Uni-D messaging just before the Cataclysm. I'm told they are strong believers."

"They sure are," Yolanda added. "It's where it really bedded down for me. Being this far out from anywhere and the solace it provided inspired me. We'll find friends here for sure." A smile was shared amongst Olisa and those near to her.

The COO took control of the room. "Enough chatter. It's time to depart. Second group, Leena should be back here in thirty or so minutes."

Hames, Robero, Francois and Siggy watched the others step through the pressure door to board the shuttle. Once it was sealed shut the remaining crew heard a large metallic clunk. From the porthole on the pressure door they watched the others in the shuttle drift away from the ship. It looked as if the far wing tip of the *Medens* had broken away from the ship. The triangular craft drifted for a moment until its exterior electrified and fizzed with life to reshape itself. A more aerodynamic trianglular structure formed with viewing windows, a new protective coating and four holes at its back where a pulse engine materialised. Power exerted itself from the newly constructed engine points and the floating shuttle took a turn and dove into the atmosphere of New Galapagos.

As they watched from a side window, Hames thought it looked more like a flat stone skipping over a lake than a transport vehicle for his colleagues. Thirty minutes later, as promised, Leena was back with the shuttle. It re-morphed then connected again to the *Medens*. Coming back through the pressure door, the first pilot looked greener than usual, more like her Mermain cousin.

"Not a problem," were her first words, followed by a guttural burp.

"Argh, I am not looking forward to this." Robero had a pained look on his face.

Leena patted him on the back as she led him into the small craft. "Sick bags are in the shuttle. Help him out, Francois."

Once they were strapped in, Leena set them on their way. "Okay, you lot. It's not a joke, it will be a tough ride." With a clank behind them Hames could hear the *Medens* release them to the world below. The shuttle once again fizzed and changed itself ready to attack the atmosphere. Initially, there was no disturbance apart from the odd shake and movement in the craft. As they dropped further below Hames could see the slight glare of a heat shield activating and then nothing as they penetrated the thin upper atmosphere of New Galapagos.

"Right, once we pass the sun-ray it will get nasty," Leena called back to her passengers. "You can have a good look at it – I've put the filters up so it won't blind you." Hames and the others could see outside the left window the large orange and yellow burst of energy, captured in the sky by a ring of red. For the moment they flew past the circle of light imprinted on the sky they sat in silence and stared at its Sol-like appearance.

Francois was shocked. "It is unnatural to fly this close to the sun."

"Well, it's not a real sun is it?" chastised Leena.

Robero breathed in deeply and for a second looked an improved shade of grey. "I will say it may not be real but it is making me feel better already." The shuttle took a turn and moved further downwards through misty layers of the New Galapagos atmosphere. Finally, they could see the planetoid. Rocky ruins and mountains first came into view, followed by dense green jungles and bright blue rivers. The focus of their journey then began to narrow in on a grouping of buildings. Hundreds of homes and structures straddled cobblestone roads, all leading up to a large area, wider than the *Medens* itself and circled by a tall white wall. Standing imposingly in the centre of this space was the biotower. It stood like a tall white tree that had entwined itself around the the rocky mountain.

Leena had seen it before. "I reckon it's like a giant root, sprouting alongside and into the rocks next to it. It travels upwards so naturally."

"It's not natural, I can tell you that," responded Hames. "At least forty storeys high."

"It is stunning though." Francois gripped tightly as the shuttle shook. "Look at the front section, around two-thirds up. The gold lines – I think that's the symbol of Universal Direction."

Leena adjusted the shuttle for the turbulence. "You're right, it's the Uni-D symbol. It's damn impressive when you get a bit closer; it's massive."

"It is truly wonderful to see the connection we have with these people," said Robero as he gulped back on his sickly throat.

The shuttle shook and Siggy clumsily flipped through some pages on his data-pad. "Actually, the biotower is where we are going to meet the Administrators and it's where we'll stay for our time here. It's basically a one-stop shop in terms of how New Galapagos works. Although the planet is covered in jungles and forests you notice these don't reach into outlying buildings of the citadel. The biotower creates some sort of field that keeps the oxygen in and limits the vegetation growth that could be propagated by the sun-ray."

"Where does the water come from? I cannot see any lakes or evidence of precipitation." Robero's dark sunken eyes widened as they searched the planteiod from their high viewing point.

The shuttle shook some more and Siggy had to resort to shouting back at his colleagues. "There's a load of underground aquifers that the citadel accesses. Oh, one other thing, the biotower keeps out all the nasties. All the bugs, the viruses, some small animals and so on. You gotta remember, it's not called New Galapagos for nothing. Lots of strange animals out there. Hey, Robero, you might get to try some of them, I'm told the fuchsia snake-rat tastes like chicken!"

"Now you're just trying to make me feel sick. You are not a funny man, Mr Conson."

The shuttle took a sharp turn, moving to ride a different cloud channel. This one appeared to take them away from the main populated area.

"Leena, you do know where you are taking us don't you?" Francois clung tightly to his seat, a hint of fear edging his normally peaceful tone.

"Yeah, yeah, it's just the way these currents work."

Siggy called over to the large and nervous Cernunnosian man, "Don't worry, Francois! Have some faith in the route down to the surface. We'll soon reach the landing pads of the biotower."

"I have faith in the route, just not this tiny metal box; it is going to be the death of me!"

"Weren't you a stealth trooper back on Cernunnos?" challenged Leena as she again took the shuttle into a steep turn.

Francois shook all over. "No coffins of metal and circuits to crash in. Being alone in space is like swimming in the calm of nigh..." A sudden jolt sent Francois' eyes closed, a visible indication that he was done talking. Fortunately for the five of them the journey was nearly complete and not long after they landed on a large pentagon-shaped platform situated at the top of the biotower. From the bottom of their flattened pyramid-shaped craft four legs moulded outwards to support its arrival.

As they exited the shuttle Francois looked up disappointingly at the sun-ray. "It's not as bright as I hoped it would be."

For Hames, the familiar lighting of New Galapagos had not changed. "It's a false illumination. It doesn't get brighter even as the day goes on."

Robero looked happy enough with the dimmed conditions. "It feels good, more like night time to me."

"Yep. I always reckoned that being here was like being stuck in that twilight hour where you can see the stars and it's dark, yet there's enough light to make you think it's still daytime," said Hames.

Approaching them was what Hames assumed to be a group of representatives from the Mons Administration. Leading the group was a relatively small man with angular features, clear, green eyes and long, lank black hair. His loose tan tunic was partly obscured by a white cape, marking him out as an official of New Galapagos. Following behind was the rest of the *Medens* crew.

The small man half bowed to them, causing the official-looking cape to fall around his chest. "Greetings, I am Quartermaster Chirila."

Siggy took a step forward. "Quartermaster Chirila, please meet the Chief of the Repair Mission Team on the *Medens*, Hames Naughton."

The clothes and features of the New Galapagos people reminded Hames very much of the old South American cultures, just in a more formalised way. Perhaps the dense jungle nature of New Galapagos had given rise to this warm weather look with its sleeveless material and leather strapping adorning their bodies. He could never work out whether this was fashion or function for New Galapagoans. Now he had returned after such a long time he could see that it was probably both.

"Ah, Chief Hames, you may not remember me. I was a young supplies administrator when you were here last. I tended the store where you would obtain equipment for studying the particle debris."

"I'm sorry, Quartermaster Chirila, I'm a little older and can't recall you. Please, don't be offended. I've met many people in my time and thirty years ago I was preoccupied with fixing the Cataclysm."

"It seems you are still working on that problem." While this point dug into Hames he decided diplomacy was best so he ignored it. The Quartermaster turned to the rest of the guests from the *Medens*. "If you would all come with me please, a brief interlude before we move down to the welcome ceremony at the base of the tower." The Quartermaster took them across the platform and through a small entry point leading down into the tower. They wound through a long white corridor before arriving at a blank white room where the nine-member delegation from the *Medens* sat through what Yolanda likened to a 'bio-induction'. In this process an imprint of their 'bio-specs' was taken and studied, albeit briefly, to tell if they were a threat to the community and to identify them throughout their stay on the planetoid.

Emmalyne was perturbed. "This is ridiculous. I studied up on this place. It's full of so many exotic bugs it's more likely to be a threat to us!"

"Protocols, honey, you gotta love 'em, and live with 'em," Yolanda whispered, trying to quieten her crewmate.

Once the crew was completely scanned, they followed the Quartermaster to a lift to join the welcome ceremony waiting far below. There were no doors to the lift to shield them. The lift drove straight to the ground level of the tower, arriving at a wide portal to a large open area covered in charcoal-grey cobblestones. People were everywhere, some coming and going, but mostly congregating around a large group of tables with bountiful dishes of meat and fruit. Quartermaster Chirila moved forward and the group followed.

Francois breathed out, his stress evaporating. "Is this for real? I think I have found that we truly are part of everything. Look at all this food and the people!"

It could have been a scene at any Earth celebration with people of different sizes, appearances and colours mingling with each other. The only point of consistency was the sleeveless tunics with wide leather strapping. Mostly these were tan-coloured with slight variations. Like the communicator bands of the crew of the *Medens*, the Universal Direction symbol glowed brightly in the adorned leather straps.

New Galapagos did appear to have a very mixed culture and intermingled heritage. The space was so wide and full that Hames believed there could be hundreds of people there waiting to see the crew from the *Medens*.

Francois pulled at the white short-sleeved tunic that held his bulging chest. "Emmalyne, remember what the analysis said of the temperature on Sol-Concio? It is the same all year round. I think that its the same from the sun-ray."

Emmalyne squinted up at the sun-ray as it sat high above the biotower. "It's pleasant enough. I don't know...there's something weird about it."

As they walked towards the dais Leena gave some level of agreement. "I'm not exactly sure what you're feeling but the atmosphere isn't natural. The heat is minimal yet the whole place feels so stifling."

Emmalyne tucked a small stray hair from her ponytail behind her ear and murmured, "All of them, even the young ones, their skin is drawn and slightly worn. It's like they're slowly cooked over time."

"I'd keep that one to yourself for the time being. Not the best relationship-building comment!" urged Leena.

"Hmm, agreed."

In the final steps of their approach to the dais, the Quartermaster spoke to Melini Guerra. "Yourself as leader of the *Medens* will stand first, here below the dais, in line with your team for presentation. Tar-Mons will give no speech; your arrival has been communicated to all. He will simply lead the welcome ritual as I highlighted to your ship's manager.

The COO frowned slightly at Siggy. "Sure, not a problem."

Hames could see Tar-Mons walking up onto the dais and taking a central position. He was tall and thin with grey hair, and carried himself rigidly, only his glowing tanned skin suggesting health. This man was the lead administrator for the planetoid system of New Galapagos. His clothing was similar to those around him, however, Tar-Mons' attire carried a gold trim on all elements of the strapping and clothes. On his back, large and bold, the symbol of Universal Direction glowed and spun intermittently.

Tar-Mons raised his arms above his shoulders and the crowd went silent. Following this signal the Quartermaster lined up the *Medens* crew and, instructed by Siggy, they held out their left hands. Each one of the crew was provided with the anticipated small flying tortoise-like creature, the chelonoida. They all clasped their hands over their small creatures to stop them flying away.

"I think mine is sweating," whispered Robero. He gave Emmalyne a look as if to say 'I'm going to be sick'. Emmalyne replied with a half-snarl, half-smile on her face as if to say, 'I'm tough, watch this'.

"Welcome to New Galapagos!" exclaimed Tar-Mons, who proceeded to squash the chelonoida bug together and then tear its shell coverings away. With grace and ceremony the tall leader drank the liquid remaining in the cup-like shape that was formerly a chelonoida.

One by one each of the crew of the *Medens* slammed their right hand down on their turtle bug and squashed it completely. Leena, despite nearly fainting, held onto herself; but ultimately it was Robero who failed, turning and running for a nearby green-leaf plant. To finish the ritual, an attendant came by and washed each of the crew's hands with a dark yellow towel taken from a golden metallic bowl. Hames breathed in the heavy air as he waited for his hand to be cleaned.

After the ceremony, the *Medens* crew all lingered in the wide-open space and exchanged pleasantries with the residents of New Galapagos. By all accounts many had come out to meet them. There were the expected questions: 'What was it like on Earth?' 'Who is going to succeed Legeles de Nord?' 'Would that harm Universal Direction?' There was genuine concern for this latter question. It was quickly apparent to Hames that although it was a farflung colony, Universal Direction had prospered on New Galapagos. It was a colony in space, somewhat lost from the other colonies and the home planet.

One particular man with a shaved head and stubborn cleft chin furiously engaged him on the topic. "Tell me, Chief Hames, what are the day-to-day conventions of Uni-D on Earth? Here we start the day with meditation. Every day all of New Galapagos takes part in one conjoined thought." The man interlocked his hands as he spoke at Hames. "It is truly amazing, the silence, the calm. One day I did not participate and it left me disturbed until the afternoon; at that time I corrected the absence of peace. Is it the same on Earth?" asked the bald-headed man again.

Hames was at a loss as to what to say. The earnest conviction of the man was tangible, but he had no answer. "Um, ar, it's not something that I can really talk about."

At that moment Olisa walked past. "Ah, please meet the ship's healer, Olisa. She can inform you of Earth's Uni-D practices much better than I." Before either Olisa and the man could raise an objection he left the conversation.

Hames was anxious to reach his quarters. Thirty years ago he was not ready to scan the skies for clues to Nisha's fate. Now, the biotower above him stood tall and massive in the sky, holding details about ships travelling along the former PCD route. He hoped there could be something here, some new information in his search for answers.

Hames found Siggy talking to the Quartermaster. "Sig, I'm tired. Is there any chance either of you could set me up with my quarters so I can get some rest?"

"Sure, Chief," said Siggy as he pulled out a datapad and laid it down on a nearby wooden bench that hovered above the stone ground. The screen lying flat on the wood resembled a piece of paper. After a moment Siggy turned back saying, "There you go, I've sent the details of your room and which wing it's in."

"Chief Hames, let me take you to the guard over here. He will take you into the tower." The Quartermaster gently placed his hand on Hames' back, setting him on his way towards a far corner of the open concourse that held bending arches and beams that drove into the ground. In his quiet the Quartermaster tone sought assistance from a man in the customary tan garb who was armed with a simple stick.

"Guard, please show Chief Hames to the Blue Wing."

"Yes, Quartermaster."

Hames followed the guard as they moved through arches and beams that criss-crossed behind them. In a few short steps they had left the gathering, passing through the winding corridors that obscured any identifiable pathway. Eventually they came to a cavernous opening that reached all the way to the top of the tower. An endless path twisted its way upwards around the walls, interlaced with many arch-crossed entrance points, similar to the one he had just came through. "I'd forgotten about the grand staircase; it's quite spectacular."

"Have you been here before?" asked the guard."

"I came here a long time ago although I never had access to the main citadel buildings. I was just a scientist working on the outskirts of the village. Right after the Cataclysm the Administrators were pretty careful about engaging with people."

The guard gave a customary overview of the grand staircase as they twisted their way upwards. "The staircase is the fulcrum of the biotower. The building has been built and grown outwards from the rock face of the dolerite cliffs behind it. One third of the tower is located within the rock wall itself, containing various rooms and services that go towards maintaining the biotower's functions. Other front-facing sections form the elements of the tower that the Mons Family administrate from, such as the hall and offices below, all the way up to the various landing pads."

The sound of Hames' footsteps rang out on the misshaped stone pathway. "I always thought it was like a hundred-metre high rocket had crashed into the cliffs."

The guard seemed unimpressed at this casual interpretation and continued on with the official service and directions to Hames' accommodation.

They walked in silence until they reached Hames' room, Blue 1304. He thanked the guard and turned to see his room was essentially open to all. There was no door, just the stream of arches and beams crossing over and over each other. He wondered how this worked in practice, this persistent openness in their lives. Before moving onwards, he looked around and made a guess that his room was quite close to the rock face. This assumption proved correct as the back wall for this room was completely dolerite. Layered in shape and colour, the grey and purple textures formed a jagged escarpment.

Hames stood still, facing the sharp rock wall. He felt excited after waiting so long to get back to New Galapagos. In recent years he had begun to hope that the biotower may help his search for Nisha. The imposing view of the dark dolerite held him a moment more before he progressed to the nearest terminal. He commenced his calculations and accessed the scanners of the biotower. He investigated the local space of New Galapagos and its moons, searching for Nisha's ship, the *Python*. Soon he would be able to tell if she had been here or at least travelled along the Mermai Tentacle after the Cataclysm.

The screen sounded a tone of completion and Hames felt nervous and excited. There were references to the *Python*. Butterflies flew inside him like they were on fire. These flames, however, were quickly distinguished. The findings were from trips long before she was lost to him. The space and debris around the New Galapagos planetoid showed no signs that the *Python* passed through New Galapagos during the Cataclysm. He could see the *Python* had travelled through the tentacle many times, just not when Nisha could have been on board.

He considered whether his calculations were wrong, but his visceral disappointment told him they were not. All those times she had been on Earth with him. Performing such a mundane job at the shuttle docks. He had been studying, day and night when they could have been together.

He went to turn the screen off but was surprised to find it showed an additional reading, one that should not have been present. Somewhere in the space near New Galapagos, somewhere very close to New Galapagos' surface, there was a faint trail of particle debris that was new. Much younger than since the Cataclysm. It was so faint he could not tell if it was thirty or five years old. This dim trail gave him hope and curiousity.

A wave of tiredness came over him and Hames dropped his head to the pillow. He placed his hands on his face and exhaled. It was like the finishing of a race that he had been running for some time. As he drifted off to sleep Hames knew that the race was not done; it was just the first leg of a longer journey.

Chapter 18

Mons Administration

EARLY IN THE MORNING, VARIOUS representatives of the biotower entered his quarters through the open archways that served as the gateway to his room. From porters to the quartermaster, he'd had so many visitors checking on him that Hames decided to relocate to a public space for breakfast – it could feel no less private than his current situation.

He walked down the winding pathway that hugged the inside channel of the tower. Some storeys below he came to a circular opening to the skies, presenting a view that showed most of the citadel. At the bottom of the biotower he saw a long stretch of cobbled road surrounded by rivers of rhyolite and rudite that stopped at a cluster of small brown and green buildings. The spectacular view extended to take in the grey rocky hills that provided a contrast to the forest-like colours of the New Galapagos residential areas. Standing at this opening, Hames felt the tempered morning warmth from the unnatural illumination source, the sun-ray. This strange device gave a feeling of limited life to New Galapagos. There was air to breathe and humidity to foster densely deep green vegetation; yet, the strength in the cobblestone roads that led to the citadel belied the fragile nature of a community that sat within the inhospitable outcroppings of the gaseous planetoid.

He found the *Medens* crew sitting at a large rectangular table in an adjoining room. All except for Francois, Olisa and Robero were to attend a first liaison meeting with the Mons Administrators. Hames took a seat amongst his colleagues and quickly caught up on the conversation.

"The Mons Administrators were in fact just a family, like any other. The family Mons." Siggy looked proud to impart this knowledge. "The grandfather of Tar-Mons, Jon-Mons, was the first project leader when they started building the biotower. As time passed the family began to take on a more direct leadership role given the importance of the biotower."

"That's right," confirmed Yolanda. "When Hames and I got here after the Cataclysm the biotower was finished. I remember there were still some inhabitants making the the transition from protective dome biospheres to living under the atmosphere of the tower and the sun-ray."

Siggy pulled a tablet out in front of him. Flicking through pages of data he continued. "By the time of the Cataclysm it was Jas-Mons who was leading the colony. It says that after the Cataclysm, the Mons family retained the role of administrators of the colony. Over time it seems that they came to run the place by virtue of setting up the administrative agenda."

"No elections?" asked Emmalyne. "Not even representative ones?"

"Nothing about elections here." Siggy scrutinised the data. "Early on there wasn't enough power or communications available to make participatory government viable. Hell, according to this update they only just got molecular tech in the last twenty years. The Mons family seems to run a pretty tight ship when it comes to letting the community get a hold of new tools."

"Sounds more like authoritarian rule to me," suggested Leena, grabbing a small purple fruit, her small hands splitting its skin open like a mandarin.

The COO firmly put down a tall brown mug of peppermint tea. "Careful, cousin. We're visitors here." Melini Guerra then addressed the remainder of the crew. "It's about time we headed off to the meeting. Finish up, the quartermaster should be here any moment now."

Siggy and Yolanda gulped down the remains of some sort of bread substance while Hames took some water.

"This meeting is important." The COO stared at them all. "The *Medens* needs new parts, energy cells and more. We can raise all this with Tar-Mons without all the formal ceremony 'clap-trap' as Yolanda put it to me last night. Remember, we just have to get the supplies and take some time for the crew to relax. That's all. I don't want you to do anything additional in the time you spend with the New Galapagos people. Nothing more than what you need to do." The look accompanying her final oratory suggested this was not a request.

Quartermaster Chirila entered and addressed them in his customary quiet tone. "Good morning, all. I am going to take you up to the biotower administrator suite, which is sixteen floors above us in the top quadrant."

"Okay, thank you." The COO's response was strong and clear.

Yolanda could not help but notice that the volume of her leader's reply made the quartermaster wince. "Quartermaster Chirila, is everything alright?"

"You all may have noticed that we speak at a level that can be heard, yet not spoken of."

At this point something clicked for Leena. "The lack of doors and hushed voices – you're very strong practitioners of Universal Direction are you not?"

"That is correct."

Leena paused as she pushed on the golden spot just below her chin. "You have no doors so you are open to all, yet still need privacy so you speak quietly. To be heard by the person with you but not by others who might share your secrets."

"Well done, young woman, you are most observant!" The quartermaster was beaming at Leena through his greying teeth. "If you would all follow me we'll progress towards the Administrators offices."

They all followed the quartermaster up the long pathway that served as the main route in the tower. Hames walked next to Leena, barely able to match steps with the young woman, who appeared stronger in legs and more equipped for their trajectory. "I'm impressed you saw the dynamics of how the people in this building operate."

Leena slowed her short step to match Hames'. "I was curious and, to be honest, uncomfortable last night sleeping in my quarters with no doors. I accidentally walked straight into Melini's room and then Yolanda's. It's a bit strange."

Hames breathed out hard realising he was struggling with this long walk to the upper quadrant. "They're a very trusting society. Even more so than ours. This is what Universal Direction has brought them – sleepless nights."

"I guess they have nothing to fear from each other. They live the connection that is life. It's a literal open-door policy."

Hames scanned the entrance points and people they walked past as they moved onwards. "It's not without mitigations though. Guards everywhere, speaking in hushed tones. Perhaps having no doors is more symbolic than anything else."

"Maybe. They all seem to have slept really well. No dark eyes like all of us. They truly don't feel like anyone or anything is going to harm them. I'm not sure the same can be said of the *Medens*. I don't feel safe anymore on board. I actively fear for our safety." Leena was not joking with Hames and he was in no doubt what she was suggesting.

"Let's see how Dao goes fixing the repair engine, see what else we find. Maybe we can make the *Medens* safe again and keep on with the mission."

Leena seemed to accept this begrudgingly. He was grateful then that she changed the discussion.

"Robero's still no good. I left him in the room. He's limited to only a few different positions." She opened her hand and started to count these off. "Staring out the window, lying in bed, or curled up."

Emmalyne moved up next to Hames and Leena. "You forgot one."

"Did I?"

"Delivery of violent sickness into the basin."

"Oh yes, that one. Not sure how I forgot. Thanks."

"Anytime."

The group deviated from the winding pathway, moving inwards through another maze-like corridor and into a room about the size of their mess hall. Large royal blue tiles lined the floor of the wide room, the uneven triangles and hexagons overpowering the low-ceilinged room. High white poles of all different widths and lengths bent through each other, crossing together to form a high cathedral roof that thinned out the closer they got to the edge of the tower. Tar-Mons waited for them at the end of the room at the far balcony. Moving alongside him they could see an image of New Galapagos even more expansive than the one from their breakfast viewing.

Stepping stiffly in his gold trimmed tan tunic, Tar-Mons motioned for the group to a set of padded midnight-blue benches that formed a large circular area in the cavernous meeting room. Hames realised this oval shape presented a very similar arrangement to what was in place on the *Medens* in the quadrangle. They all took a space on one of the benches and faced towards the thin, upright lead administrator.

Tar-Mons spoke as others had done in the biotower, with a strong yet quiet voice that carried to them all but no further. "I welcome you all to New Galapagos and trust you slept well on your first night?"

"We did thank you," responded Melini. "Some of us ate too much; that's what happens when you're on rations for a few weeks."

"I would assume so. COO Guerra, we are most happy to have you here. The chance to interact with outsiders does not come often to New Galapagos. It is a strange thing to have a connection to places so far away yet not be able to visit them."

Melini responded stony-eyed. "That would be challenging. You seem to have become very self-sufficient out here."

Tar-Mons shifted slightly on his seat, his tall body ramrod straight as he sat with his hands resting on his knees. "I believe we are in a great place. A safe place of single purpose. Before the biotower was built, New Galapagos was a dome settlement and a precarious one at that. Our community had fought on for years, determined to not let man's first interstellar home become its first interstellar ghost town. In our confined domes, our society hit a point where it couldn't expand and this removed the ability for people to grow and learn." Tar-Mons sat forward and looked at them all; they in turn gave him the attention his position accorded him.

"You have to understand that without the ability to grow, a society begins to obtain a tacit feeling that there is no future to be had. We had a truly beautiful outlook with dense jungle and forestation pushing out from the bottoms of our rocky mountains. This visage was just that – a picture not to be touched. In some ways, I am sure this irony gave weight to the depression that began to set in. Some of the original settlers became affected by those dark feelings but felt compelled to stay." Tar-Mons stood up and looked to the open balcony. "Some ended up taking their own lives. That is what self-sufficient can look like."

Yolanda leant in from her seat. "I never reflected on that part of your history. You had just gotten rid of the domes when I first came here. I only thought of the technological advancement and not the other factors behind it."

Tar-Mons spoke again, "Leaving our dome city behind was one milestone, but the future still necessitated action. So heavy in carbon dioxide is our cloudy planetoid that the atmosphere needed extensive adjustment. Of course, we also needed energy as without this our growth aspirations could not be realised."

Yolanda led the conversation with Tar-Mons. "There were only a few carbon farmers here the last time I came."

"And many more now. It is the most common activity on New Galapagos and the most beneficial. We built collectors and converters that helped us reduce the temperature on our planet and increase the oxygen content. It allowed us to move out of the domes and settle underneath the protection of the biotower. Our primary challenge now is one of continued heating, and we have recently commenced a new energy process that gives us balance in our ecosystem. And I am proud to say my son conceived this idea. Before, we used a process whereby we simply sequestered the carbon dioxide we collected. We now take this through a closed cycle conversion process that collects the carbon and reforms it into a viscous matter. This is then combusted to produce energy. The by-product is then captured and the process repeated."

"Old school," Leena commented.

Tar-Mons was confused. "I'm sorry?"

"Don't worry, Father, she is just saying the combustion technology we use is a little barbaric." A young man approached, his footsteps loud on the blue tile floor. He was not overly tall, but thin, with tanned skin, showing he had spent most of his life under the New Galapagos sun-ray. His long brown hair was full and curled to his shoulders. Hames' initial impression was that this created a sense of freedom about him. A freedom and energy that seemed to glow from his bright green eyes.

Leena was clearly taken aback by the objection to her comment from the young man; her distinctive dark-skinned nose pointed downwards in slight embarrassment. Hames was sure she was not expecting the competition of youth in her niggling of Tar-Mons. "Perhaps you could rub two sticks together instead of burning fossil fuel?"

The young man looked back at her and stopped, looking thoughtful. "You know, I think I could make that work!" he said, laughing. "No, it sounds too simple and where's the fun in that?"

"All of you, please meet my son, Bas-Mons. He is the one who came up with the 'barbaric' idea of energy derived from carbon." Tar-Mons was not blind to the conversation between Bas-Mons and Leena. "We are a carbon-intense planetoid so we do need to manage the atmosphere. The alternative is being driven back into the domes."

"Pleased to meet you all." Bas-Mons sat on a padded blue bench beside his father, crossing one leg over the other, patting the sole of his foot in an energetic fashion.

"Thank you Bas, you may join us." Tar-mons turned to the crew of the *Medens* Well, that is part of the story of New Galapagos. I am always happy to share this with esteemed visitors."

"Thanks for sharing, it was fascinating," Melini Guerra commented while leaning forward holding her hands together.

"Given your presence enriches us, my question is what can my Administration do for the *Medens* and its crew?" Tar-Mons' face was bland. Hames wondered if the question was born of caution or perhaps even protection.

"We need resources, parts and maybe even people. We need to fix our repair engine and our light alignment drive," the COO explained.

"Yes, I have been told of this. It is my understanding that you have incurred two significant accidents of late on the *Medens*, the latter costing you the lives of two crew members."

"That's correct." The COO was uncomfortable at Tar-Mons' awareness of the *Meden's* recent history. "Pending our own investigations and repairs of the ship, it's our intent to complete our mission. To repair the Mermai Tentacle and make contact with Mermai."

"My dear woman, Mermai is twenty-three light years from Earth and another twenty from here. It is unreachable. A ship cannot last that long. In any case, you will be very aged by that time."

The COO remained solid in voice and body. "The *Medens* is built for this and I've got all the time in the world to give to the trip. We may only make it a few more years before turning back – we'll see what happens. To do this, though, we need some supplies and parts or we have no hope."

"You are welcome to help yourself to food, clothing as well as parts from our space fleet. It is retired and I have no need for ships any longer. The days of travelling to distant worlds is over. We look at what is in front of us, not above or behind us."

"Thank you."

"Please liaise with my quartermaster; he will help you with supplies. As for people, I will say your chance of convincing someone to join your impossible mission is as likely as its completion. You are free to ask."

"Thank you again."

Tar-Mons stood up straight and tall, and moved to the balcony that revealed the world he oversaw. He looked back at them all, his body now looking tense under its tan cloth and leather straps. "I do ask that nothing is done to disturb the balance of my community. You are welcome here, but we are not accustomed to outsiders."

Yolanda looked curious, scratching her short grey-cropped hair. "You've Universal Direction here. It's clearly so throughout the tower. Where could there be imbalance and disharmony?"

The quartermaster stepped in for his leader. "Tar-Mons mentioned the presence of carbon farmers earlier, yes?" The small lank-haired man received nods of recognition. "Carbon farming is essential to the system of life on our planetoid. The growth of the carbon farmer numbers and community brought with it a cost. The last influx of carbon farmers came just before the Cataclysm, when Jas-Mons was lead administrator. We learnt that they brought a belief system with them that was contrary to Universal Direction – Montaxe-Deus is what it was called. This small group grew till it became a problem. It became secessionist in nature."

The COO looked puzzled. "Montaxe-Deus was around for a short period near the time of the Cataclysm, but it's been dead for as long as I can remember. For nearly forty years."

Tar-Mons spoke over his quartermaster before he could answer. "Whether it was this Montaxe-Deus or another group. A passionate and contrarian resolve against Universal Direction was present on New Galapagos."

Hames remembered Montaxe-Deus vividly. He had been with Nisha's cousin Jasiah at its end. The idea that it was somehow present on New Galapagos made no sense to him. The movement had dissipated; he even remembered the editorial from the Agency communication channel about its cessation – 'Not with a bang but a whimper' – which wasn't entirely correct. Now here was this administrator of a farflung colony telling him they still existed. The idea felt so foreign it was like being told a long ago part of history had been transported to the present, like pirate ships or the Quadrant priesthood was back.

Tar-Mons continued. "There was a challenging moment about thirty-two years ago which made the then Administration focus much more on what was happening in the outer parts of our community." Tar-Mons spoke with such conviction that Hames himself felt as concerned as the Administrator. "There was an attack on the central village, in particular the Curate-mente centres. They never got very far; though; we stopped them after the first building was destroyed." Tar-Mons looked proud at this moment.

Hames then started to feel uneasy about where this conversation was leading. He had a history of being unhappy with Curate-mente, particularly on New Galapagos. He noticed both the quartermaster and Tar-Mons eyeing him with more regularity than the others at this point.

The COO herself bore a serious look of apprehension. "I hadn't heard of this; I don't believe anyone on Earth has."

"It's something that we have sought to keep to ourselves. We believed they wanted to either secede or become infamous."

"Oh my mind," said Emmalyne, "what was the damage like?"

The quartermaster answered the request for details. "Horrific – the bombs were big. Not dirty, yet powerful enough to destroy the Curate-mente centre and everything else within a fifty-metre radius. The perpetrators were caught and sent away, banished to space. The remnants of the group were then identified and pacified over time so they would no longer be a problem. As of today, we know of the descendants of these people as some live further out from the citadel. We believe them to be active solely in their minds, if that.

Tar-Mons returned to his warning. "We have reached balance and harmony on New Galapagos. It is true, we are all everything, together. Yet your presence may raise questions, so for this reason you must be aware."

Leena revisited her concern for the safety of the crew. "Oh, I get it. Now we're in mortal danger."

"Ha!" Bas Mons could not contain himself. "That sounds almost exciting but lacking some reality. Most of us take a tolerant approach; if people want to cling to old ideologies then we should be accepting of their presence."

"Thank you, Bas," cut in Tar-Mons. "We also try to encourage them to make different choices."

"Just what do you mean by encourage?" asked Leena.

"What my crew member is asking is how did you filter out the people my crew needs to be aware of?" The COO translated her cousin's difficult question.

"We offer all people on New Galapagos meditation sessions, re-education and Curate-mente."

Hames was curious. "Surely if they were Montaxe-Deus they didn't take up this offer?"

"Some did, most didn't. There was some opposition to it due to an earlier incident with a Curate-mente centre." Tar-Mons was focused on Hames.

Hames had been dreading this moment for a long time; the trepidation he had felt when relaying stories of New Galapagos were about to be realised. He addressed Tar-Mons and the quartermaster, "Okay let's get on with it."

The quartermaster turned to Hames. "So I did identify you correctly yesterday? You are Hames Naughton who came to New Galapagos over thirty-five years ago?"

"Yes. That's me."

"What's this about, Chief?" Emmalyne's brow furrowed.

"Yes, Chief, what is this about?" the COO demanded, pulling her hair back to reveal frustrated brown eyes and a tight green face.

Tar-Mons saved Hames from answering. "I will leave it to Chief Hames to explain to you another time. Suffice to say, in the years prior to the Montaxe-Deus uprising, there was an accident with one of the first ever Curate-mente sessions conducted here on New Galapagos. The first session was an ambiguous failure and there was an angry protest."

"And?" asked the COO.

Hames froze; he felt his body lock, although he was sure no-one expected him to be moving anyway. He was about to share something quite personal and felt the eyes of his crew members looking towards him. He looked at his hands; the broken compass on his right seemed to move despite a lack of command. After a moment of silence he confirmed the truth. "I was one of the angry protestors."

Silence fell across the open liaison room until a hushed frustrated tone came from Melini Guerra. "Well, that's just great!" The COO stood up angrily and walked around the room.

Hames continued. "A young member of my team, he was the participant in the first Curate-mente session. He died."

The quartermaster was shocked. "Oh, Chief Hames. You have it wrong. Marak Chakla is still alive today."

"What? I saw him go! I saw them switch off his support!"

"That is true," the quartermaster confirmed. "However, he continued to breathe and live. He has been that way ever since. His family consented to continue his support. Now, Marak Chakla is housed in the hospital level here in the tower."

The turn of events had left Hames stunned. A moment ago he was a quiet participant in the meeting. Now the whole point of focus was on him. "Can I see him?" Hames looked at the still slightly angry-faced COO.

"Sure, but take someone with you," responded his leader.

Emmalyne moved towards Hames and spoke to him quietly. "I can see where your head is at, Chief. Let me come with you." The COO nodded in agreement to this offer.

Hames looked to Tar-mons. "Is that okay?"

"Please, make arrangements with Quartermaster Chirlia to guide you to the hospital level. It is acceptable to me."

"Thank you." Hames was genuinely appreciative of the flexibility offered by the Mons Lead Administrator. As had been the case during the last month, there was another puzzle to solve that did not involve Nisha, another distraction.

The meeting continued, although most of the participants looked outwardly stunned at Hames' revelation. After they had gone over a range of logistical matters with the quartermaster the gathering was over. They all left Tar-Mons' rooms and a collective feeling of confusion seemed to emerge.

Yolanda voiced their concerns. "That was not the discussion I had been expecting. It should have been something like 'be wary of this fruit' or 'stay away from that area'."

Leena put forward a pragmatic Universal Direction spin on the matter. "What will be, will be."

They all began to wind their way back down the inner pathway of the biotower. As they walked, Siggy gave them some credits and sent them on their way. Hames stepped away from his crew when he reached the deviation point towards his quarters. He was conflicted; normally he was single-minded about finding Nisha, or at least what had happened to her. Now he had a barrage of issues and challenges to address. Hames would soon see Marak Chakla, who he'd long thought dead. He entered his quarters and once again felt alone and unsupported. This time almost shunned by the crew. At least there was Emmalyne; both of them were somewhat broken and out of place but they were in it together.

Chapter 19

Montaxe in Mordon

THE WEAK AFTERNOON SUN FEATHERED the terraced houses in Wandsworth Town, giving Nisha a feeling of finality in the day. In truth, she sensed it was only just beginning. Apprehensively, she walked up Marcus Street, away from Jasiah's flat, to approach Sean Watson, her follower from the Agency. Light glared at her as the sun reached the horizon; she put a hand up to to cover her eyes. Nisha processed the situation as she walked towards the Agency man. Her experiences over the last week had opened her eyes and triggered a growth in her confidence. Maybe Sean Watson had been watching her or maybe even protecting her. She still wasn't sure which. There should be nothing to worry about as she had done nothing wrong. Although, she did feel a slight discomfort after sitting in with Healer Milar earlier that day. The Curate-mente session at the Universal Direction clinic had been interesting but disturbing.

Reaching Sean Watson, who now stood clear on the footpath for all to see, she wiped away all her concerns and spoke assuredly. "You're not very good at being conspicuous, are you now?"

The man from the Agency looked as stiff and bureaucratic as when they'd first met. Awkwardly, he played with the white shirt cuffs under his plain charcoal suit. "Miss Pinto, if I wanted to go unseen you would feel entirely alone."

"Oh, really."

Sean Watson looked unconcerned. "The Agency has now classed you as a non-threatening individual. You're nearly clear on that front."

Nisha's intuition had been sharpened of late. She was getting much better at reading between the lines. "So is this about Jasiah?"

"In his flat, did you see what has been occupying his time?"

Uncertain of this government man, she decided caution was required.

"Not really; and it shouldn't be of concern to the Agency anyway."

"It is a concern, Miss Pinto. We know what is happening. We just want to know if you know, so you can help him." Sean Watson looked uneasy. "Jasiah is about to lie down with dogs. Dogs with a plan."

Nisha laughed and then regained her composure. "That sounds like a good move to me. Is he in danger?"

"That depends on your definition."

"Is he in danger? Tell me now or I'll be right back at my pod deleting all the files on new PCD destinations that Hames has identified."

Her threat was fake and insincere, merely meant to motivate the man from the Agency. Sean Watson acquiesced. "Fine, he's not in immediate danger. Not at all. For some reason, Jasiah has become obsessed by the bombing. He's following and publicly interacting with a group that we think may have caused it."

Nisha moved to the end point of the discussion in a deflated tone. "You mean Montaxe-Deus."

"So you do know what is going on?"

"I know there is a meeting tonight in South Morden." She felt sheepish now as she was sharing information about her cousin with someone she didn't fully trust. "Just because there is a meeting on tonight doesn't mean Jasiah is attending or involved or that this group is bad."

"Just as we've had a person watching over you so too has someone been watching over Jasiah."

"Oh."

"Right now, someone is following him. Jasiah is travelling the route to Morden."

"Well, I wouldn't advise they try to get into the meeting with Jasiah. It said bio-tag invites only."

"That's right, we can't attend." Sean Watson looked directly back at Nisha and raised his eyebrows ever so slightly, enough to present a question to her.

"You think I can go, don't you?"

"It had crossed my mind."

"I'm not Jasiah you know! The bioscanners will read my DNA straight away and realise this."

"Maybe, maybe not. The tech is still fairly new and your Mermain DNA may confuse the meeting parameters."

"That's a big maybe."

"And in any case, if you are picked up you have a great cover story. You're there to find and join your cousin."

"What is this group anyway?"

Sean Watson turned around and leant against a low red-brick wall. "Montaxe-Deus are an organisation presenting themselves to disaffected groups as a peak body representative."

"Disaffected groups?" Nisha nervously adjusted the mermaid necklace on her chest.

"All sorts of groups. Religious ones – any faith or spiritually related practice that fears its subjugation by Universal Direction." The man from the Agency then held out his hand towards Jasiah's flat. "And, Montaxe-Deus is attracting some of the people from the other worlds, like Mermai. These people have some sort of twisted fear inside them that harmony across the colonies is a conspiracy to take back control their home worlds only so recently won."

Nisha shook her head and paced on the footpath. This was not what she thought her cousin would be taken in by. "How has this all come about? Universal Direction seems so new."

"The work and ideas around Universal Direction have been in progress for some time. The Cetana Initiative has been around for years. The Paris and Cape Town Conferences were landmarks in the work on consciousness and its structures. We've assumed Montaxe-Deus has been formed in response to this work. What we can best determine is they've pulled a lot of groups together who are already disempowered. These groups are so fearful of the future that they are willing to work together under the same banner. Hence the name 'Montaxe-Deus', which loosely translates to 'assembly of the gods'.

Nisha stopped in her pacing, her voice almost folding in on itself. "What does Jasiah want with them? What do they want with him?"

"Well, they want membership. Anyone who feels lost or disenfranchised with the direction Earth is heading qualifies." Sean

Watson's voice was professional and sounded genuine. "We're not really sure of their true intent. We strongly suspect, though, that it was Montaxe-Deus behind the bombing at St Paul's."

Nisha's mind started to race through the possibilities as she rubbed her sore shoulder. She felt in shock and pushed a muscle in hard to trigger a return to reality; she barely felt it. "They could be a really dangerous bunch. The blast at St Paul's, it wasn't pleasant."

"We don't know for sure it was them, that's why we need more information. They're naturally suspicious and some of their members have covert agency backgrounds. They would spot us a mile away."

"I'll go, for sure." The words came out of Nisha's mouth before she understood what she had said. Her pulse began to race, her body innately alert to the situation ahead of her.

Sean Watson stuttered for the first time in Nisha's company. "I thought I might have to persuade. I can't promise you that there won't be danger. Most likely not, they will just want to scan you to record your presence."

"I understand." What was she thinking? This was madness and Hames would be furious. No need for the Agency to ask her not to say anything, she was going to keep this to herself.

Sean Watson pulled a communicator from his pocket. "Excellent, but you don't have much time. I have a transport up above that will run you down to Tooting Broadway station. You can grab the last few stops before you get to Morden. The Montaxe-Deus meeting starts in forty minutes."

The transport came down; a smaller version of the one that took her to Dover every day for work. It unfolded, allowing the two of them to board, and in a short time they were at Tooting Broadway station. Nisha rushed through the gates and her ticket was paid with a scan of her bag. It was only after she finally boarded the train and took a seat that Hames again came to her mind. He would definitely be unhappy with her. But she had to go. Jasiah was her only family on Earth and he was entering something that sounded wrong.

The train arrived and Nisha alighted at Morden station. She wondered if the Montaxe-Deus session was the destination for any of her fellow passengers. She walked down Aberconway Road then found herself in Morden Hall Park, heading towards Morden Hall. The sky above was a fading blue, an afterthought of a sun removed from the day. People walked past her in ever-increasing layers of clothing; she shivered slightly in her flimsy navy jacket and pants. Morden Hall stood clear in the park, centuries old and white. She realised some walkers in front of her were indeed focused on the same destination. That there was only a handful of people made her worried. Would she be found to be a non-invited person?

At the entrance, two open bay doors were parted and framed by a temporary bioscanner. She was going to have to pass through this if she was to find Jasiah. As she approached the doorway her body filled with so much tension and anxiety that she couldn't act naturally. She did not know these people or what they were doing. *I'm really crazy! What am I doing!* she thought to herself, *Turn around now!* Against her own wishes, Nisha pushed on to the entry point. When she walked through the gateway the pace and pressure of her heart beat warned that fainting was her next step. This nightmare was realised as someone next to the bioscanner called to her.

"Excuse me, excuse me! Can I speak to you for a moment?" A tall man with short blond hair and plain yellow and green clothes beckoned her over.

No guns or alarms were blaring; she reasoned this was still a recoverable situation. "Sure, not a problem."

"You're not on our list here. But I can tell from the scan you're Jasiah's relative. Are you happy to consent to that?"

Nisha opened her face up. "Um, yep, that's me."

"Thank you. We welcome people from Mermai, a beautiful place. Jasiah is already here and the meeting has just started so please grab a spot where you can."

Nisha's relief at her successful entry was such that she nearly found herself dancing forwards into the large meeting room. The hall was imposing with light wooden panelling reaching up to a high ceiling. It was even colder inside than the park. The room was in distinct disrepair, the wooden walls and linoleum floor worn and faded. In the middle was a square dais with chairs flowing back from each side for four rows. The groups of old rusted seats with rough red cushions were sporadically filled with attendees. The meeting was certainly not a fully subscribed affair neither was it situated in a decadent setting.

A young dark-skinned man was walking around the small dais in the middle. Nisha thought him to be only twenty years old, give or take a year. In contrast to his obvious youth he carried himself confidently. Nisha could not tell if he was Indian or from another part of west Asia. He could have possibly been Mermain. The youth was talking passionately to the group and frowned at Nisha when she entered. This forced her to grab a seat before she could spot Jasiah.

She checked the room as she listened intently. The young man spoke clearly for all to hear. His tone was persuasive and familiar. "They say Universal Direction means no harm to us. Tell me, when have we let our government give us instruction on what to believe in? First it will be an example, then a guideline and then, a law." Mumblings of agreement flowed around the sparsely populated room. "It won't be long before they remove the last of your churches to make way for more healing centres." More disagreeable mutterings followed. "Or remove you from your colony and make you a citizen of the Universe!"

Nisha finally spotted Jasiah, to the far left of the same row as her. He was clapping his hands at this last rousing call from the main speaker. Jasiah saw her, too, and his eyes turned from surprise to blazing fury, then almost embarrassment.

The young man leading the meeting walked slowly around the sparsely populated chairs. "It won't be long before they start taxing your fundraisers or banning the use of your native tongue and languages." The small group of men and women of all shapes, sizes, colours and creeds began to rock in their old rickety metal chairs.

Nisha looked over at Jasiah, who glared back at her with frustrated eyes. She turned her head downwards so nobody could see her and mouthed at him, "Why are you here?"

His response was clear. "Go away." Nisha noticed that Jasiah was wearing a white wristband. She had briefly noted it when she first saw him and now as she looked around the assembly, she saw that others were wearing this too. The wristband was a rope as thick as her thumb and was not tied or bound together by anything obvious. The last two ends of the rope protruded back out and away from the wrist for a few centimetres.

"As your Montaxe-Deus counsellor, I can represent you, and I can tell you that representation is necessary." The counsellor moved around the square dais. "You are all here from different faiths and followings. Here from different worlds and systems. You don't need to forget what you are or move on from this. You are complete as you are."

The fever seemed to be cooling. "Let's all take some time together. This is for you. This is your choice. You can think about anything you like. Say anything you like. Your beliefs are yours alone to keep now and for time eternal." The counsellor moved to one of the front rows. He sat down slowly, closed his eyes and laid his palms open in his lap.

Nisha took this as her moment and stood up. In her row, people were choosing different methods to experience their 'moment together'. Some moved their legs up onto the seat and crossed them. Some took a rug from their arms and moved to the front of the room and began saying prayers. Others sat still and silent. As she walked past two people kneeling in her row she realised that all of the people present seemed harmonious in their choice and accepting of the others around them. There were no darting eyes or words of displeasure. In the mixture of faith and race around her there was only peace. It seemed remarkably similar to Universal Direction in that regard.

She reached Jasiah, who was sitting with his eyes closed. His head twitched at her arrival, but he didn't acknowledge her. A poke to his side brought him to life.

Jasiah reacted with a firm whisper. "Bugger off, Nisha. If I wanted you to join me I would have asked you!"

"I don't understand what you are doing here, you're not religious!"

"This isn't religion! No-one should be told what to believe."

"Who is doing that?" she asked.

"Legeles de Nord and Universal Direction are doing that. They're telling people to be harmonious and forget the past. They'll be telling people from Mermai to forget about home soon enough!"

"The Eagleton Accord has given the colonies self-determination; they can't take that away. Anyway, Uni-D is just an idea for harmony, it's not a law or anything."

"Yet!" he snapped back at her, loud enough to finally disturb one of the kneeling attendees in the row behind them. He quietened again. "And don't get me started on the healing centres, I've read about those."

Nisha had to admit he had a point there. The healing centre had felt wrong to her, although it still seemed to be a free choice. "Righto, I get what you are saying. This just doesn't seem like you, Jas. You don't have a diary let alone an agenda. Why are you spending your time with these people?"

"Why shouldn't I take some time with these people? You're leaving here in a week or so, what should I be doing?"

It hurt but it was true. She was leaving and though they had come together her plans had totally revolved around Hames and getting back home to Mermai. She hadn't considered Jasiah. He'd always been on a path of exploring the Universe whereas she was here just to be with Hames.

"You could come home!" whispered Nisha, trying not to shout. Calming down again, she persisted. "You could visit the other colonies: Cernunnos, Earth 2, Lugus. Take a trip to Mars for all that's worth!"

"I like Earth, it's my home now."

"You're kidding me! You just stay in your room and paint pictures of Mermai! Come home!" Her voice was loud now and people meditating four rows in front of them were lifting their heads.

Jasiah looked at the ground before responding. "The blast in front of St Paul's. I wasn't hurt a bit. At first I was scared, but nothing touched me. I stood up and looked around at people on the ground or running for their lives. Some were just standing there screaming. I was looking around for you and then I just found myself in the middle of it all. Something clicked inside me. Afterwards I kept researching Uni-D and what it was. It doesn't feel right, Nisha. Reading up about who else feels like me is how I found Montaxe-Deus."

"You're still in shock. These people are lonely chaps with too much time on their hands. Just come home with me and it will all be okay," she pleaded.

"No, you go home. Here is where I can best protect Mermai. Where I can save our home from Universal Direction!"

"Jasiah..."

She was cut off by the voice of the counsellor, who began to reiterate his message of representation and protection of the past. "And finally, as many of you know, we are looking to set up more assemblies for Montaxe-Deus, in and around London, and the rest of Earth and her vestiges. Your assistance is vital to our success. Whether this be financial or through volunteering. Now we have finished this assembly, please come and join me and we can plan our coming activities."

"Go home, Nisha, to your apartment or to Mermai. That is your path; my path is here." Jasiah got up and joined the people crowding around the dais. Nisha watched everyone in the meeting move to the front and surround the counsellor. She could make out various notes and papers being circulated. Jasiah approached the group, playing with the white rope around his wrist. He gave Nisha one last look and disappeared into the throng. Nisha got up and left Morden Hall. Passing through the park she was a cold and lonely figure. She returned to Morden station and boarded the train with a feeling of disconnection from Jasiah, in person and in life.

The next morning, she found herself back at work for the first time in many days. She stared vaguely through the window behind her desk as the waves below lashed the white powder cliffs of Dover. A reminder rang on the screen behind her, signalling that new inductees were waiting for her on the top-level departure platforms. Taking the shuttle officers for training up around the particle control drive ships would be a welcome distraction from the challenges of late.

The lift doors opened to what Nisha always thought was the most beautiful part of the docks buildings. Underneath the large spider-shaped departure platform were glass and white steel departure lounges. Curving upward convexly, it was like a giant lid had been placed on top of eight flowers in full bloom. Light seemed to penetrate everywhere, even at its centre. The curved walls glistened but still allowed a clear view back across the cliff tops or out to the Channel.

Nisha walked past the first of the departure gates, its security scanner in full operation as a crowd of people made their way up to the shuttle. She did not need her inside knowledge of the scheduling or the departure sign to tell her these people were travelling to Earth 2. Their clothes were Earth-like, efficient and drab; the homogenous business suit. She groaned silently to herself. "Argh. Earth-2, so boring, so practical."

On reaching the second gate she could see the four inductees waiting for her. Three women and a man sat at a long white plastic bench, all talking eagerly amongst each other. Without interruption from the security scanner, Nisha strode into Departure Gate Two. "Hello all, I believe this is your last induction session before you become fully accredited shuttle docks officers?" Nods and recognition confirmed to Nisha she had the correct group. "Okay, let's get up to the shuttle and then the PCD ships."

At her final words the whole glass partitioned room locked and started to move upwards. In a few seconds the barriers of the room sunk below and they were free to board the simple shuttle craft. They took seats against the top part of the shuttle, strapped in tight, facing each other in the circular craft. The calmness of the trip belied the power and strength of the ascending shuttle. She took this quiet moment to ask them about themselves. After greeting the first three, Nisha spoke to the final inductee. "And what is your name?"

"Olive Sassonet."

"Pleased to meet you. Can I ask where you come from, somewhere on Earth?" "The east coast of the Northern Americas," responded the woman, who was only slightly older than Nisha.

"Oh wow, I haven't been there. My partner tells me it's quite an organised community, very structured and reliable."

"Yes, it is; too much so at times. I'm happy to try Europe for a time."

"Okay then, welcome aboard. Speaking of boarding, it's soon time to arrive and head to Departure Point Four, the Mermai Gate."

One of the inductees closest to a window described their arrival. "Hey all, I can see the PCD Hub. It really does look like a giant spider in space with the big body and small arms for docking rings."

Nisha was already testing them. "How big are the docking rings?"

Another of the inductees, a red-haired man responded. "They're only sixteen metres long and about as wide. They take their gravity from the main body of the PCD Hub, which is always spinning."

"Great, what else can you tell me, Olive?"

"The shuttles, like the one we're in, will land underneath the short round docking arms, and passengers then move directly to their gate, although they can access the main body of the station for food and water."

The shuttle locked into the departure ring for Mermai and the inductees followed Nisha. An officer opened the door for them and they all made their way out to the bottom section of the round departure ring.

Nisha closed the air hatch behind them, sealing off the shuttle. "Turn around please. Can you see anything wrong?"

One of the inductees called out, "The access hatch isn't locked; it shows clearly on the panel."

Nisha was matter-of-fact in her next question. "And what would you do about it?"

Olive Sassonet spoke up. "That's not our job, we're not on duty."

Another inductee agreed. "The officer and the arriving captain of the ship have a duty to lock the gate."

Nisha spoke calmly. "When you are here or down in our building you always have a duty of care to safeguard people's lives. Do you choose to leave the hatch unlocked, maybe for a child to stray through, or, do you choose to go beyond your role to safeguard life?"

The inductees all agreed that safeguarding life was a far better choice and then followed Nisha up the escalator that wound its way around the inside of the curved docking ring.

The noise of activity grew louder and more definitive as they reached the departure gate for the particle conversion drive ship. They looked back over the huge expanse that was the main concourse area, full of transiting passengers.

The red-haired inductee looked confused. "Why only five gates?"

"There are five here which service departures for most of the colonies. Two other are for commercial purposes and one for Sol-Concio travellers."

"Why does Sol-Concio get its own gate and route?" asked an inductee.

"It's a private route, not open to the public."

"Yes, but why?" asked Olive.

Nisha looked ruefully at them. "There's a special arrangement between the Agency and Sol-Concio. I don't know much more than that myself. They're not sick or infected, or anything like that. I've met them a few times now; they look normal. Okay, enough of that, come this way." Nisha ushered the inductees over to the observation glass to watch the soon departing ship. "Today we're here to watch the *Python*, one of the larger PCD ships, leave for Mermai on its second run."

"How many runs do they do per day?"

"Four on most days, each way." Nisha walked to a wall nearby. "Here's a map tracking all the ships across all the routes to the main colonies. You can see all the ships active on the routes to the colonies at any one time. Of course, the Sol-Concio route is blank given its private status."

Nisha turned to the inductees. "Have any of you ever been on a PCD ship before?"

The red-haired male raised his hand. "I have. It was a family holiday to Mermai."

"Great! My home! What did you think?"

"It was okay, but I got sick. Too much spicy meat. Also, the trip was weird, like we were here then at Mermai in about a minute; it made my head spin."

Nisha felt protective of her home. "Maybe it wasn't the meat; maybe you were feeling a little sick from the trip. That can happen."

"Oh," was the only response.

Out of the corner of her eye, Nisha could see some movement outside. "Look, there's the *Python* moving off its docking bay. It's going to drift outwards about four hundred metres and then, gone!" The group turned and watched the large, flat, triangular-shaped ship slowly disconnect from the main hub and float to a point where it looked far smaller.

Olive pressed her face against the glass. "It's so flat and thin! From here it looks two-dimensional."

Then, as promised, the ship simply vanished.

"Isn't that amazing?" stated Nisha.

"I guess so," offered the red-haired former PCD traveller.

Nisha ignored the lack of enthusiasm. "What did you think, Olive?"

"It seems okay. I'm not really sure if it matters that much to me. I'm just happy to have a job in London, far away from America. Once the Cetana Initiative held its Dualism Forums in Washington I knew I had to come closer to the Agency and the steps towards liberating consciousness."

Nisha was taken aback and not ready to find a person so committed to Universal Direction on her crew. "Hmm. Sure, it's good to have a job, Olivia. Oh sorry, I mean Olive."

The other inductees laughed and one offered Nisha some consolation. "Don't worry about it, mam, we've been getting her name wrong all week. It's been a running joke for every new manager we meet."

"It's true," confirmed Olive Sassonet. "My name is a bit old fashioned. Most people who know me just combine my name. Please, call me Olisa."

Nisha returned this gesture with a smile. "That's much better to me. Welcome aboard, Olisa, we're lucky to have you. Do try to get a bit more excited about the magic of space travel. It might help when you work in this area day in and day out."

Watching the *Python* depart reminded Nisha that in two weeks she too would be leaving Earth. For good, she hoped, or at least for a very long time. Her mind was chaotic, filled with the events at the healing centre and the Montaxe-Deus session in Morden.

Somewhere down below her two groups were emerging, reaching outwards and upwards into what she now saw as a vacant space. Since the breakaway of the colonies and the establishment of independence, it had been well documented by various editorials that Earth lacked a purpose and identity. Humans had gone to the stars and for what? Only for her people to leave and become disconnected from the home world.

It was sad, as the planet was so blue and inviting. Seeing it as a whole gave her clarity about the coming of Universal Direction and Montaxe-Deus. People down below felt sick. They felt a lack of the social nutrients that made up a happy and healthy society. Like a sense of belonging or a sense of collective purpose. Uni-D and Montaxe-Deus weren't born out of nothing; they are here to fill the void. Nisha shook her head. *Earth is broken and it's going to get worse.* She hoped Hames had started packing; it was time to go home.

Chapter 20

Marak Chakla

EMMALYNE AND HAMES STOOD ON the observation deck towards the back of the biotower, waiting for Quartermaster Chirila. White short-sleeved overalls clung to their muscles, barely flapping in the slight breeze. The quartermaster was to show them to the hospital wing that housed Marak Chakla, Hames' former colleague of over thirty years ago. They spent the moment surveying the valleys ahead of them. The formations looked deep and foreboding, covered in a dense blanket of cloud and gas.

"It's so peaceful here," Emmalyne began. "The cloud doesn't move; like it's asleep waiting for the day to begin."

Hames cleared his throat. He was still apprehensive since the morning's meeting with the Mons Administrators. "I know what you mean. The light from the sun-ray only partly stretches out this way. So the outer areas of New Galapagos aren't very well illuminated."

"I've read about the formations over there. The Impenetrable Valleys they're called. Tall and impregnable with only the smallest of openings between cliff faces allowing entry and exit to their landscape," said Emmalyne

"I didn't know that. You do read up, don't you?" A slight wind came down from the valleys and blew through Hames' bead and hair.

"Well, I wouldn't be out here if I wasn't a curious personality, now would I?"

"No, I guess not."

Hames motioned to a far right point of the skyline. "Do you know what that station is in the mountains?"

"The one with the beam shooting up to the sky? I'm pretty sure it's the other half of the sun-ray. Providing the circular holding pattern for the Sol-like image."

"Tell me, as you're a curious person, what do you think has been going on with the *Medens* in the last month? First, there was the accident in the LAD turret room, and then the disaster with the repair engine."

"I'm really not sure. It's been amazing to see the world out here, be on the *Medens*, to see the Ugor Eels. It's also been pretty terrible. I still feel guilty about losing Xavi and Chung. If we'd worked quicker, got the particle drive venting sooner, they might have lived."

"There wasn't anything more we could have done. The gravity well was off, the alarms, you know. Look, the ship was doing everything we thought it'd never do."

They both sighed, contemplating the vista of clouded crags and crevasses. To Hames, the air seemed fresher up here than down on the cobblestones that surrounded the base of the biotower. He began to talk through his internal concerns. "I've never heard of a LAD having an accident with its viewing turret. Never before have I seen a repair engine fail like that. For both to happen within weeks of each other... can that be a coincidence?"

"Maybe. The accident with the light alignment viewing turret didn't kill anyone and a repair engine has never been in service as long as the one on the *Medens*."

Hames stretched an arm back to scratch his head. "You were pretty hurt by the accident in the LAD turret room."

"I didn't enjoy it, sure. In truth, it wasn't exactly an incident that would cause the ship to blow up. There was a burst of power followed by some extreme light. That's all. The repair engine reversing? That'd be an incredibly complex feat to engineer, especially when there are many other ways to blow up the *Medens*."

Hames appreciated Emmalyne's logic; it would be difficult to engineer the repair engine failure. Maybe a coincidence was all it was. "You might have it. I think we can read too much into all of this. Perhaps losing someone is all we need to think about."

Emmalyne tipped her shoulder into his. "Maybe you can't explain loss, maybe you just have to feel it, accept it and move on."

Hames leant further forward onto the balcony, his compass-bearing hands gripped tight on the white metal bar. "Easier said than done."

Emmalyne leant back while holding the rail, stretching her arms before again coming back upright. "That's true., I'm not sure I could cope with much more. I was feeling really bad before. Olisa has been great. She helps me relax and feel calm. Somehow, though, I want more. I want to feel more settled."

Hames laughed. "You're talking about mind-healing to the man who got kicked out of New Galapagos for protesting about it."

"What happened, Chief? You can tell me now the COO and the others are off seeing the citadel."

"I'd reached New Galapagos about four years after the Cataclysm. Like the *Medens*, it was a light alignment drive ship, just without the repair engine work and maintenance to slow us down. I had a theory to test: whether we could launch a particle conversion drive ship away from Earth, maybe in a less damaged area. Unfortunately, it was all a mess. We tried a few test ships but the PCD debris covered most of the local space, right up to the fourth moon. There was no clear path through to Mermai or another colony. My last hope was to move further out from New Galapagos space. So we set up camp on that last moon, you know the one we passed under when we cruised towards New Galapagos."

Emmalyne held tight against the rail as a small gust of wind caught her long blonde hair. "Oh yeah, I remember. The Verona Moon, the one with all the high cliffs."

"Yep, that's right, the one Yolanda mentioned. I had a team there mapping pathways and testing the space around the moon. The goal was to find a clean pathway to Mermai, no matter how small. If we'd been successful I might have even jumped on one of the test ships to Mermai myself."

"I think the COO would have done the same if she'd been there."

"Maybe. It was inhospitable work, though. We set up some temporary shelters, because the fourth moon is much like Earth's - low gravity, no atmosphere. Anyway, while we were there we found something. A reading of energy emanating from a cavern near one of the sites where we set up our telescopes. Two of us went to investigate it and we took a carrier with us, a young man we enlisted from New Galapagos to help out."

"What's a carrier?"

"It was basically a person to do all the stuff the scientists didn't want to, like, um, carry stuff around, energy packs, supplies."

"Oh."

"Yeah, the equipment was really heavy, you know."

"Sure."

"Like I was saying, we all suited up and went into the cavern. A few metres into this eight-metre-wide opening we found what looked like a pool of water that had a shimmer to it."

Emmalyne peered back at him intensely. "It sounds sort of beautiful."

"You have to understand that we were in suits. The atmosphere was almost non-existent; a real barren moon. So to find this stuff was amazing. What we didn't realise was that it stretched behind a set of large boulders. I'm never really sure of what happened next but the carrier stepped back, I think, put his foot in the water and he took a shock."

"He was frightened?"

"No, like an electric shock. I just remember a faint green flash behind me and I turned to see him falling to the ground. For a while we all just stood there like fools not knowing what to do."

"What happened to him?" Emmalyne asked

"He was out cold. His heart slowed to almost a standstill. We took the shuttle back to New Galapagos and moved him straight to a med-centre in the citadel."

Emmalyne leant against the balcony. "Okay, I'm guessing that the carrier is the guy Quartermaster Chirlia mentioned this morning, Marak Chakla?"

"All of my team were upset when it happened. After we got back we left him with his parents who stayed by his bed. A few days went by and he was coming in and out of consciousness. I went to see him on the fourth day and he started making no sense at all, really random stuff. The names of his friends, objects and shapes – it was like his mind couldn't focus. That day he sat up straight in bed and grabbed my left hand, the one with the good compass, and pointed at it with a grimace on his face." At this Hames held the hand up and waved it somewhat. "He then laid himself back down. The next day, though, he seemed to be better; he looked to be coming out of it even though he was still groggy. He was starting to communicate more calmly and normally. The boy couldn't remember the week before, but he seemed fairly normal."

"So what went wrong?"

"Marak's parents were new converts to Universal Direction and the local healer from the Uni-D centre had also been visiting him that week. The healer convinced them to sign-off on Marak undergoing mind-healing. I didn't know this, and in any case I wouldn't have had a say. I came back the next day and saw the aftermath. Marak was there in body but gone in mind. No talking, no real brain activity. He wasn't there anymore."

"A few of his friends and I met up at the local inn." Hames looked regretful. "We got too drunk. We were just very angry. Later in the evening we headed to the mind-healing centre to look for answers. Some of us started banging on the doors. It got rowdy and finally security came and escorted us away."

"Chief, you must have been pretty upset."

"Marak was a good kid. He didn't deserve what happened to him. He just wanted a job and to help us out. It wasn't even much of an adventure. The thing is, the next day news spread of what happened and somehow it came out that we caused Marak's condition. Someone spread the word that it was the trip to the moon that had caused him to become comatose. I knew differently, though; he was coming good before the Curate-mente."

"Maybe, Chief, you don't know everything that happened. There sounds like a few variables at play."

"My head tells me you're right, my intuition says different. I didn't get much time to investigate it. A man from the Administration told me Marak had passed away once his life support was switched off. Not long after that I was asked to leave New Galapagos. No-one here wanted to look at fixing particle debris or plan for safe PCD travel in the future."

"I'm sorry your research was cut short; it must have been tough having another door closed on the chance to find Nisha."

"I knew at the time it was a dead-end. I took all the data away with me. Later on when more Ugor Eels started appearing, I realised there might be some connection between the eels and what happened to Marak. I checked the readings from one of the tools that recorded the pool of shimmering water that Marak had stepped into. The reading was similar to the egiya readings used to identify Ugor Eels. Not exactly the same, but definitely comparable molecular structures. I never thought to come back and check, it just wasn't on my mind."

At that moment Quartermaster Chirila arrived to collect them. "Ah yes, the Impenetrable Valleys in the Outer Zones." The Administrator stepped forward to join Hames and Emmalyne. The quartermaster's white, green-trimmed cape flapped behind him. "They are beautiful and will stay that way eternally. The population of New Galapagos will never reach a level that requires an invasion of such an ominous space. Did either of you get a chance to visit the citadel today?"

"I did for a short time. It was very real and alive, especially after being on the *Medens*," Emmalyne said.

"Well, that is a different comment, thank you," the quartermaster replied and then turned to Hames. "Are you ready to visit the patient?"

"You say that like he's still alive."

"That's because he is, Chief Hames. Although he has not been awake for many years, he is very much living. Come; you can see for yourself."

Hames, Emmalyne and Quartermaster Chirila wound their way through the back corridors of the biotower. They found themselves following a line of complex catacombs that reminded Hames of the *Medens* and its network of passageways. The rounded white plaster walls were identical save for the presence of a picture, a sign or an entry point at every bend or length.

"I'm completely lost, Chief," Emmalyne complained out loud.

"Me too."

"Where's this guy taking us? It can't be to cut out a kidney, he could just print one out!"

The quartermaster turned around and looked in horror at Emmalyne. "Ms Biggs! I can assure you that you are in no danger whatsoever here."

"Oh," said Emmalyne, sounding almost disappointed. She mouthed 'sorry' to Hames, who was looking wide-eyed and incredulously at her.

The quartermaster continued to lead them on through the indistinct route. "I must say, Chief Hames, it's strange that after all these years you are one of the only people to return to New Galapagos and on the first day back you are confronted with the reason for your departure."

Begrudgingly, Hames participated in the personal interrogation that was dressed up as small talk. "I guess it's some strange form of escalated coincidence."

As if reading his mind, Emmalyne looked over at him and said, "There are no coincidences, Chief."

"Indeed!" exclaimed the quartermaster.

Emmalyne winked at Hames and he felt a warmth inside that replaced the anxious coldness he felt. He believed that it was the right thing to do to see this man who had been practically lying in state for thirty-odd years.

Finally, Hames, Emmalyne and the quartermaster climbed a narrow spiral staircase that wound its way upwards into a small circlular area. Surrounding them was another room, a large circular room with a glass floor-to-ceiling wall along the length of the wider circle. They could see one continuous circular room with a bedding area, food preparation, medical supplies and other machines. In one section, lying completely still on a hovering platform-bed, was Marak Chakla. Feeding tubes pushed out of his arms; his frail, diminished body lay underneath a light yellow plastic sheet. Hames looked down at his feet and on the floor of the small room that held him and Emmalyne was the Universal Direction symbol. Its yellow and orange colours were faded and lacked the luminescence of other representations Hames had seen before.

"If this is a hospital I want to get sick right now," said Emmalyne.

"Really, is this going to be appropriate, Chief Hames?" The quartermaster was clearly concerned about Emmalyne's uninhibited language.

Hames pulled the quartermaster aside. "It will be fine. Leave it with me, I promise it won't be a problem."

"As you request." The quartermaster groaned as he fidgeted with the white cape resting behind him.

"How do we get into Marak's room, Chief? For a guy who carried bags and has been in a coma for decades there is a lot of security," said Emmalyne.

"Can I ask why you just didn't put him into cryo?" asked Hames.

"Resuming him after cryogenic storage would most likely terminate his existence. It was not worth the risk. In contrast to his lack of physical activity, Marak Chakla has grown to become important over the past few decades. He is a symbol of our commitment to Universal Direction. His life is a symbol of our presence in the Universe, that it is forever unending." The quartermaster recited two of the Universal Direction conventions to close his views: "The past is past; we are all here together."

Hames and Emmalyne walked slowly around the glass circle together. Hames couldn't help thinking that the quartermaster was taking some joy in this display of Uni-D's connectedness and it infuriated him to see it. Marak should have been left to pass on ages ago. Starving his comatose body of life support seemed preferable to the years of restricted existence that had rendered Marak almost unrecognisable to Hames.

The quartermaster moved towards the exit stairs. "The entry point is in front of the food preparation area. I have programmed the wall to let you through." A part of the glass shifted away to allow them entry. "If you don't mind, I will leave you to whatever you think will come of this. I will be waiting down the corridor in the last entry point we passed before our arrival."

"Okay. I think can find you." Hames looked at Emmalyne, who gave a shrug.

The quartermaster made his way down the staircase and the two *Medens* officers approached the entry point in front of the food preparation area, accessed the circular room, and walked around to Marak Chakla's bed.

Hames was perplexed. "I'm not sure this will take long or what I'm really doing here. Where do I start?"

"Why by saying hello, I would think." Emmalyne's smile made this all the easier for him.

He approached the bed and looked down at Marak Chakla. He was breathing normally and perfectly still except for the movement of his chest up and down. He was not paralysed, but not properly working either. He looked terribly different to when Hames had last seen him – gaunt and drawn with greying long hair. Hames finally mustered the courage to talk to him. "I'm sorry this happened to you. I don't know what I could have done. I'm just sorry it ended up like this."

"That's nice, Chief, really nice."

They both stood there a while longer, still and silent in Marak Chakla's personal ward. Hames felt tired and rubbed his eyes. As he did so he thought he saw a twitch in Marak Chakla's face, just on the eye lid. "Is that normal?" he asked Emmalyne.

"I don't know, what do you mean? A person to lie like this for years on end?"

"No, his eye twitched. People in comas do that sort of thing, yes?"

"Not my area of expertise, Chief. Maybe, I think I've heard of that happening." Emmalyne held up her hands in confusion.

They continued to stand at either side of Marak Chakla's bed, the silver hover platform contrasting against the gold-like colour of his bedding.

"I swear I saw it happen. When I rubbed my eyes." Hames felt energised by this. "Do you think he knows I'm here, that we're here?"

"Chief, they said he's been like this for decades; I'd say not."

Hames stood there feeling as if something else were to be done. This couldn't be it. He slowly reached out towards Marak's hand.

"Chief, be careful, we shouldn't."

"It will be okay, don't worry." Hames slowly took Marak Chakla's hand in his. At the same time Emmalyne moved closer to the bed to ask Hames to stop and her hands brushed the top of Marak Chakla's body. Hames felt a surge. Like the one he had felt before in the viewing room of the light alignment drive. This time no warmth, just a flow of alertness within him, taking him to the deeper parts of his mind. He felt it in the place where the disappearing image of Xavi had called out to him and Emmalyne. Beyond the echo of those final words he saw the simplest of images, a faint square in some mist or clouds. It was there, then gone. He recoiled in shock, stepping away from the bed. It had only lasted an immeasurable moment of a second, but he knew this memory would not fade.

Emmalyne was standing across from him looking amazed. Her face and eyes were tumultuous like the skies of New Galapagos. Her arms half extended asking a question her brain could not answer. "It was Xavi. He spoke Xavi in my mind or I felt it, just like when Xavi left us!"

Hames was confused. "I didn't hear anything, but I felt something. I felt an image in my head. Did you feel one too?"

"No, no." Emmalyne was frantic. "I just heard Xavi, again and again, like we had heard him before. I, I..." Emmalyne stopped mid-sentence and ran, moving fast to the opening to the circular room and down the stairs before Hames could speak. He was in two minds as to what action to take. The image he had *felt* in his mind was nothing sinister or nefarious. Finally, after some contemplation, he departed to look for Emmalyne.

He looked everywhere for her; through the winding corridors and around the nearby levels of the tower, knocking at her room and checking the main areas. She could not be found. He finally came across Bas-Mons and Leena in one of the lower levels. Bas was showing Leena some animals that clung to the inner rock face of the mountain that ensconced the bio-tower. One of the small creatures which looked like a rabbit-sized white sloth was slowly creeping up Bas Mons' arm.

"Have you seen Emmalyne!" Hames demanded.

"Yes," replied Leena. "She looked mightily upset when she came through. She said she was looking for Olisa and that it was time."

"Oh," said Hames.

Leena stared Hames down. "It's not your decision, Chief. As much as you feel a need to protect Emmalyne, she has to make her own decisions."

"It's just that..."

"Just nothing, Chief. It's about her, not about you. Don't take it personally. I think it's something she's been considering for a while. Something must have upset her to make her decide today." Leena and Bas walked back around some more rocks to restore the small creature to his home.

In that instant, Hames felt like he had decades ago; a familiar sense of loss and grievance creeping in again. Like he had lost Nisha and like he had lost the carrier on his crew, Marak Chakla, he was about to lose Emmalyne as well. She was so smart, so funny. Maybe after Curate-mente she would be the same, no different than before. *Today. Why today?* he asked himself. Then out loud, he echoed Emmalyne's words: "There are no coincidences."

Chapter 21

The Original Tortoise

IT HAD BEEN THREE DAYS SINCE the crew of the *Medens* arrived on New Galapagos. Some of the crew had been to the villages that bordered the biotower; others took in more traditional experiences, sights like the Museum of New Galapagos that showed its history way back to the first dome settlements. Hames had spent most of the second day on the *Medens*, relieving Dao and making his own fixes to the repair engine. All the requied parts had been secured and now they had to grind through the process of putting it all together. He wasn't a fan of spacewalking so he gladly left this task to his technical lead and returned to the surface for his third day.

Hames arrived back at the biotower in the afternoon and noticed the small tortoise-like bugs, the chelonoida. It was fascinating how they clung to the lights on the shuttle platform at the top of the tower. This gave him a vague recollection of a tortoise, a memory separate from that of the chelonoida.

"Are you sure about this inn place, Chief?" asked Leena as they made their way down the long winding pathways of the biotower.

"The Original Tortoise, yep, pretty sure." Hames reassured the group that was following him, Leena, Robero, Francois and Yolanda.

Robero looked worried at a repeat of the activities at the welcome ceremony. "Are we going to be drinking more chelonoida juice?"

"No, no. The Original Tortoise is an inn, a place to relax and get a drink."

"Oh, thank my mind," said a relieved Robero.

"Look, it's not too far from the biotower; it won't take long to get there."

"Will they welcome strangers, Chief?" Yolanda asked. "Remember what Melini said about taking care in the community."

Hames had to think. "I reckon so. From memory, the Original Tortoise accommodated all types – workers, administrators – anyone from New Galapagos was welcome there. It's kind of towards the outskirts of the citadel. It used to be a quiet place, we won't get noticed much there." The group made its way further down towards the bottom of the bio-tower and Hames enquired about absentees. "Is Emmalyne coming?"

Leena's voice was quiet and cautious. "No, Chief. I think she just wants to stay in the tower for the time being." She eyed Hames' attire; he was still in work gear with a brown-smudged white t-shirt and his *Medens* blue work pants. "You're not exactly dressed for going out, Chief."

"Maybe not. There should be some agricultural types; I'll fit in fine. What do you think Emmalyne and Olisa are doing?"

"I think they're working on a healing plan."

"She just needs to talk about things; she doesn't need someone manipulating her amygdala."

The small first pilot scrunched her forehead. "It's not as if you've tried it. Neither of us have."

"I guess we're both unjustified in our positions."

Leena's voice was calm as they approached the bottom of the biotower. "I don't think you always hear me, Chief, whether it's sharing the Williams Sphere or making a decision about turning back to Earth. It's not always about what you want or what you can do for the mission. The same goes for Emmalyne – it's her decision, not yours."

"You're right; I know this. I just can't help think it's the wrong decision for her."

Leena and Hames walked ahead of the others, the energy in their discussion driving them onwards. "Universal Direction tells us we're all connected. Curate-mente is something that helps that connection become real and noticeable, beyond faith and prescribed belief. I want that for myself, once Melini allows it. I also want that for Emmalyne; the thought of her feeling connected and caring for everyone around her, it excites me."

Hames reeled at this, although it was possibly justified given the events Emmalyne had endured. Coming out of stasis and feeling imbalanced was enough. The incident in the light alignment drive room, the PCD accident and the loss of Chung and Xavi had added to the stress placed on his young friend. Now their shared experience with Marak Chakla had pushed her to breaking point.

The *Medens* crew reached a large wall that also served as the front gate for the biotower. A wide space in the wall shifted and moulded to an opening at their approach. Once they'd passed through it, the wall reformed to set as the solid shield protecting the tower against the elements.

Outside the enclosure, the bustling hum of people dissipated and they found themselves in an expansive stone tile area. "It's kind of like a neutral zone out here," said Robero trying not to cough. "Only people coming to and from the tower. What a waste – it is almost the size of the *Medens*."

"And nearly eighty metres walk to the next transport point," replied Francois in his strong French-Cernunnosian accent. "I can only assume that is part of the intent. You have to walk up and admire the tower. It makes people respect it and only attend it if necessary."

It was close to evening and despite the continued warmth, the unnatural light provided by the sun-ray now dimmed. Much like most of the *Medens* crew, Leena found the artificial sun disconcerting. "I can't get my head around this tech. It feels so out of place yet it's creating something that's so necessary for us to feel, well, human."

They all turned around to look at the sun-ray station that poked out from the side of the biotower ensconced in the mountain.

Its faint beam shot up on a sharp angle into the sky, filling the waiting circular pattern.

Robero was interested in the device from an engineering point of view. "It may look like it is coming from the tower, yet if you look more closely you can see the sun-ray station is anchored in the mountain next to it. It takes structural integrity from the rock face and pushes its resonance frequency back into that mountain. The biotower couldn't handle that, it would shatter." He stopped and put his hands on his knees. "I can look high no longer, it is making me sick." A green-looking Robero dropped to his haunches.

"Ho there!" called Bas-Mons running up behind them, his arms and legs flowing in motion with his long brown hair. "My father has asked that I accompany you."

The five of them looked at each other and then back at the young man in the tan tunic. Leena glared at Hames as if to say no; however, something inside him suggested an opportunity.

"Thank you, Bas," responded Chief Hames. "You're welcome to join us, especially if, as your father suggests, our presence may have not gone unnoticed."

"I don't think there is anything to worry about," he said grinning. "Well, we can find out together, can't we?" With a wink he put one arm around Leena, who grunted, and the other around Hames. "The Original Tortoise I suppose?" François and Yolanda both nodded at the same time. "Predictable, yet a good choice to see the real New Galapagos."

Yolanda picked up Robero and they all began their way down the concourse towards four wooden hover-carts. The carts were simple platforms sitting a metre above the ground. The timber on the edge of the carts was rough and splintering, encouraging passengers to sit on worn padded cushions in the centre while they dangled legs towards the ground.

Bas-Mons eased their uncertainty. "Hop on. It's not comfortable but it's safe. They move at a snail's pace." Bas-Mons grabbed the front point of a hover-cart and jumped on. The cart stayed steady for the crew members to embark. Hames and Leena joining Bas-Mons on either side, their backs to each other. Bas-Mons began to move the cart down towards their destination using controls that Hames couldn't work out. The village came into view immediately, slightly obscured by an increased volume of chelonoida flying about them. "The houses, they're all bent trees pulled over and over each other!" Leena said in amazement.

"That's right, kid," said Yolanda. "Not much rain here and not much wind either. The wood from the Galapagoan trees is incredibly dense so they're a good material for keeping the area cool."

The cart tilted as they ran over a different type of cobblestone, larger ones that were sand-coloured. "I like to think of them as a memory of where we came from." Bas was pointing at the houses and stores. "They look like the domes from our past but are still open in spots. It's like saying this is who we are but we never again want to be closed up in a tiny environment."

Leena was uncharacteristically quietly spoken. "They all have little gardens in front of them that are so intricate and colourful. I've never seen so many rainbows in shrubs and rockeries.

"Thank you. The people of the citadel are serious gardeners. The chelonoida help as well. They pollinate most of the denser types of rock gardens."

As they drove down the cobbled pathway, Hames heard the faint ring of crystal bowls, drawing his attention to an open space between the tree-thatched abodes. He could see a large group of New Galapagoans congregating in the early aspects of a circles, all clothed in similar leather strapping and faint tan material. The further the hover-cart moved onwards the more familiar the image grew.

Hames remembered the people they met on their entry to New Galapagos. "Your whole planetoid seems quite committed to Universal Direction. I'm a little surprised that it's continued as far as it has."

"It's all I've known growing up. You know, the beauty in recognising the cosmic connection between us all."

"When I left here it didn't seem that way," said Hames.

Bas-Mons kept his focus on the path ahead of them. "Just as Earth needed a sense of harmony to sate the pain it felt after the Cataclysm so did New Galapagos."

"Doesn't that mean it's not actually real? It's just a story told to bring comfort?"

"I think even yourself, Chief, would agree that you are a conscious being." Bas recoiled in mock fear. "You're not a zombie are you?"

"No, I'm here. I just don't believe that there's a whole other form of existence that is us. That there is 'us' that we can't see. Consciousness comes from the mind."

"Not through the mind?"

"Yeah. Everything we are comes from us here." Hames pointed at his head, almost apologetically.

Leena on the other side of Bas-Mons rolled her eyes. "Don't bother, Bas; it's like trying to tell a caveman about fire."

"And what about you?" Bas-Mons asked the question to Leena's neck and then realised his error. "Sorry. I didn't mean to stare. The sentir spot is hard to miss."

"That's okay, I'm impressed you know what it is."

"We've had some information exchange from Earth, not much though. Sentirs are mainly used in the Savai branch of Universal Direction, yes?"

"Again, impressed. Let's go for three. Can you recite anything?"

"The 'individual and the all', um. No, that's all I have."

"I'll give you a pass." Leena gave an approving wink as the hover-cart took another tilt on the stone road.

"Thanks," responded Bas. "It must be hard at times being Savai, championing individual rights as part of the collective connection. Do you get in the odd argument?"

Leena looked sheepish as Hames held in a laugh. "Sometimes." She reached behind Bas-Mons' back to give Hames a small punch in the arm.

"Hey, we do things a little differently here as well. The predominate view in the citadel is panpsychism."

Leena looked more open to this idea. "Oh, really? That almost accommodates individuality."

"Kind of. We believe that most things with some form of complexity are enminded. That even the smallest of entities can have consciousness. Not just humans."

The hover-cart moved away from the view of the pending Universal Direction session to face a group of bugs and other small creatures breezing ahead of them in slow continuous swirls of red, blue and green.

Hames ducked his head as the cart passed slowly through the mass of creatures. "Let me guess, the chelonoida, those flying turtle-bug things, have consciousness?"

"Why not? Even the smallest entities like protons and electrons that spin around each other. Why can't they have some form of consciousness? Everything could be enminded."

Hames had no answer to this idea. "It's possible, but highly improbable."

Bas-Mons laughed at Hames deflected response. "Nicely handled, Chief." The young New Galapagoan then yanked the hover-cart to the left and over a bridge. "The Tortoise is halfway down this hill and over a stream."

A short stretch later, round turns and through streets with houses and businesses, the hover-cart came to a halt in front of a two-storey building whose walls, floors and ceilings were made up of multiple folded trees. After piling off the cart and through a creaky oak door, they found themselves in the Original Tortoise. A low hum of conversation was accompanied by the smell of fruity ale.

The crew found a large booth towards the back of the inn. Like the rest of the establishment it was coloured in silver and red streaks that were corniced in worn shadowy wood. Padded tortoise shell–like cushions moved beneath their backsides as they sat down and adjusted themselves.

The windows around them gave a view to what appeared to be a semi-industrial area. Francois motioned towards a large open area in the distance that didn't seem to have many buildings, just lights and fences. "Look, a small shuttle is landing. What do you think that place is?"

Yolanda lifted her tall frame above the others to scan down towards the curiosity. "Those fellas getting out. It's hard to tell from this distance – they all have bulky suits on."

"They're carbon farmers," said Bas-Mons. "It's a controlled way to bring it into the citadel. The canisters are dropped off in a utility building and taken from there for conversion."

"Do you really convert carbon to energy?" asked Francois with a deep, concerned voice. "That is very unimpressive."

Bas-Mons shrugged. "There's loads of it around the planetoid and we need to keep its content down or the planetoid heats up. The process we use to convert it keeps the balance of the amount required, no more no less."

"What's it like outside the citadel?" Leena seemed to be more tolerant of Bas-Mons as time went by.

"The biotower regulates the atmosphere around the habitable zones; outside these you can't breathe or live. There's just not enough oxygen."

"Right, time for some drinks." Hames got up and made his way to the bar and faced up to a man in a silver-and-red–trimmed preacher collar shirt with large hirsute arms. "I'd like six stouts please."

"You're from the *Medens*, right?" the publican asked

Hames looked down at his dirty top and blue work pants. "Yes, that's right, these don't really fit on New Galapagos, do they."

"Not a problem to me, mate. What's your name?"

"Hames Naughton."

"Naughton hey? I've heard that name before," said the publican. He put both hands down on the bar and leaned in. "I'm the owner of this place. Before me, my mother had it. She told a story about a Hames Naughton stirring up the place after an event on one of the moons."

In this instance, Hames decided ambiguity was the best policy. "Um, well, there was a few people and issues involved. I'm assuming I wasn't the most welcome man in town for a while." Silence persisted at the bar while the publican considered his potential customer and Hames moved onto his next tactic, flattery. "Your mother, I do remember her. A good publican and a kind woman to all. She never turned away a man looking for a quiet drink."

"That is true. There weren't as many people back then and Mother used to struggle. Now we get all the carbon farmers. Multiple shifts running all the time. The farmers work hard. So I do alright here," responded the publican.

Hames could see the Original Tortoise was indeed full of carbon farmers. Some were clad in sweaty mesh under-suits and sat draped over the bar, fatigued and burnt out. Others were obviously here for a short visit, wearing their red bulky bio-suits. About twenty or so of the carbon farmers sat around the u-shaped bar, some talking to the one next to them, some just looking into the distance, others eating bowls of red and blue leaf salads with black crumbling rings inside them.

"They come up from the landing platform down the way," said the publican, motioning towards the open area that Francois and Bas-Mons had identified before. "It's good business for the place," the publican suggested.

"I'm glad things are going well for the Tortoise. I swear I mean you no trouble as I never did your mother," Hames said, raising his hands and smiling.

The publican paused for a second before giving a rough smile. "Alright, six stouts. I'll even bring 'em over to you."

Hames moved back to the table to find Leena inspecting Robero, who looked as tired as ever. "Really, what are you doing here?"

The dark rings around Robero's eyes betrayed his tiredness. "Better to be here and shattered rather than lying in a bed not able to sleep."

Yolanda rubbed her pale crewmate's back. "Still got the insomnia?" Robero dropped his head into his arms.

In quick time the owner brought a tray with six large metal tankards of dark ale, passing them around the group. Music from what sounded like a low-sounding piccolo played against a mandolin bass as they all tried their beverage. Various reactions were elicited; Hames' was one of pleasure, Francois' one of surprised enjoyment while Yolanda offered a simple shrug of her mouth. Leena's face was that of a person who boasted they could suck on a lemon, yet couldn't achieve such a feat. Robero's expression was delayed by the continuous supping of stout until the tankard rang hollow at its return to the table. All of the group stared at him in amazement. The young engineer blinked his eyes twice; there was a clarity in them Hames had yet to see. Robero burped. "I think I need another!" The *Medens* crew laughed as their colleague rose to make his way to the bar.

Francois looked at Leena. "Not your type of drink?"

"I've never tasted real alcohol, it's disgusting. I'm not touching it anymore." Leena pushed her tankard away.

Francois took up this offer. "Well, I will not waste such a fine drop!"

"You are welcome to it." Leena pushed the cup away and looked up at Hames. "So, Chief, you were here for while. What was it like before?"

"It's no different really. Same building and colours."

"No, I meant New Galapagos and the other stuff that happened!" Leena was looking for information. It was only fair given what had been revealed by the Mons Administrators. Hames thought of himself being portrayed as a political activist; it was ridiculous. He had been more drunk than enraged when banging on the door of the healing centre that had provided Curate-mente to Marak Chakla.

"I don't know, Leena, they're old stories," he said, scratching his beard and trying not to look at Bas-Mons.

"Don't worry, Chief Hames." Bas-Mons' ever present smile narrowed to show as close to a serious side as possible. "My father and the Administration already have the story they want. You saying anything different won't change a thing, I can assure you of that."

"Hm, okay." Hames took a large drink of stout and proceeded to narrate his story of the incident on New Galapagos from years ago. "You heard some of it from the meeting with Tar-Mons. After Marak had mind-healing he was gone, like his life was drained away from him. I was angry, got drunk one night and made a racket. I then got sent back to Earth, which was fine by me." The group leaned further towards him, raising his discomfort and increasing his desire to stop speaking. "I was finished researching around the moons and wanted to investigate the Cataclysm from the original point of incidence near Earth." Hames finished this as if he was done speaking and the table went quiet, allowing just the music and consistent chatter of the bar to continue.

Leena then elbowed him in the ribs. "You never mentioned the moon story when we coasted in and watched them from the quadrangle."

"I don't know, it's not a great story. 'Hey look at this amazing moon in space we are flying under. I watched someone get electrified from some unquantifiable substance.' No, I didn't feel like sharing that," Hames said quietly.

"You're sharing now. We are all here together." Leena offered him Universal Direction with sincerity.

Robero, who was now fully awoken from his tiredness and aware of Hames' discomfort, attempted to change the topic. "What of Earth after the Cataclysm? What was it like overall? Challenging?" Robero probed Hames.

"It was horrible. When it all went down, Earth's population had a collective feeling of grief. It was polarised by the loss of its hub to the colonies. Earth felt abandoned in the Cataclysm. Some people thought the colonies had done it deliberately." Hames saw a chance to be out of the spotlight. "What about you, Yolanda?"

The chief engineer was placid in her delivery. "I feel like Uni-D helped a lot of people. The degradation of the space particles was matched by degradation on Earth of how people treated one another. It got pretty bad for a while there. Universal Direction got people back on track with a way to live life and treat others. People were committed to the idea that they were connected to each other and shouldn't hurt one another."

"You see, Chief, it's not so bad after all!" offered Francois as he downed another mug of Galapagoan stout.

"No, not all bad," said Hames. His crew members looked intently at him, half-smiling and half-nodding, as if for a brief moment he had finally seen their world. Some looked positively gleeful. This made him want to stand up and grab the table with two hands and flip it up. Spilling the tankards all over them. He visualised how they would then glare at him. Their faces would show their disdain at the emotionally unintelligent beast. He chose not to flip the table, but let them smile and hope he would see the world their way. He was disgusted at himself.

Back at the biotower, Emmalyne was most likely in a session with Olisa, preparing to wear a Curate-mente neck brace and have her pineal gland attacked by some sort of energy beam. "Listen, all, I can only come so far with you on the Universal Direction journey. You just lose me at certain points."

Yolanda understood what his issue was. "Chief, you're worried about Emmalyne, aren't you? You think she'll turn out like your friend from the moon."

"I guess so." Hames admitted.

"Both Francois and I have had mind-healing and it was fine. It was better than fine," said Yolanda.

Francois was his normal calm self. "I was a mess, Chief Hames. Cut off from my home planet and stuck on Earth. The Curate-mente I took once a year or so after the Cataclysm. It saved my life. It saved me from hurting myself and others."

Leena sat forward. "I want to try it. Melini has said yes but not until we are past New Galapagos. She says I need to be focused on piloting the *Medens*, not expanding my mind. Emmalyne, though; there is no reason to stop her. It seems like it would be good for her."

Hames got up. "I need another drink." At the bar, Hames regretted not flipping the table when he'd had a chance. He banged his hand down, inadvertently knocking over the drink of a burly man standing next to him.

The large carbon farmer in his red protection suit caught his drink, which spilled onto his hands and lap. Wiping himself clean the man was calm and not quick to anger. "Hey, mate, relax. Whatever it is it can't be that bad."

Hames held up his hands apologetically., "Sorry, just having a bad moment."

"Bad moment? Why don't you just see your healer, there's one around the corner from here." The large man pointed down the road from the inn and turned back to the bar.

This was too much for Hames. The carbon farmer had pushed the release button on emotions that he had been trying hard to contain. With the open palm of his broken compass–adorned hand, he swung his arm back and then through the top of the bar, collecting the carbon farmer's drink on the way through. The tankard was sent flying into a store of food bowls just metres away.

This in turn sent large volumes of red and blue leaves soaring across the room. Patrons everywhere recoiled in shock. Some attempted to catch knocked-over chairs and drinks. The tired carbon farmers present in the Tortoise after a long shift caught a fever. A fever initiated by Hames Naughton.

Francois and Yolanda had seen the start of the fight at the bar and had made their way across the room only to be blocked from exit due to the presence of a low internal wall. Hames, who was being held by one carbon farmer in a bear hug, looked over at them cornered against the short barricade. They were beckoning to Leena to join them. The ship's first pilot was ducking objects and bodies as she tried to make it across to Francois and Yolanda. About halfway there, a Tortoise stout was tipped down her back. The contents passed through her white jump suit and down out the bottom. At this, Leena grabbed the arm of the guilty carbon farmer as it came back over her head and threw him over towards Hames, who was knocked out of the arms of the burly farmer. This allowed the chief mission officer to breathlessly make his way to the exit. Over on the right-hand side of the bar, Robero was taking down a farmer every few seconds in smooth judo movements. For the first time since he joined the *Medens* he had awoken from his stasis illness. Robero Paldini was wide awake, standing upright and full of energy. Whether it was the adrenaline of the fight or the two tankards of Tortoise stout it did not matter, he was better.

"Robero, we are out of here!" Hames yelled over to him as Bas joined them near the exit.

"Chief, I am only just starting!"

Leena was tackling another farmer, throwing him to the back corner of the bar where they had been sitting just minutes ago. "Now, Robero!"

The fighting had slowed – some of the already tired carbon farmers lay exhausted on the floor. There was now space for them to exit the bar. Robero and Leena ran to the door that Francois held open for them. As they all exited Hames could hear the owner of the Tortoise shouting after him.

"You'll be hearing from me, Hames Naughton! The system administrator is going to hear of this!"

They all quickly jumped onto the hover-cart. It stuttered then went silent and began to slowly sink to the ground. Carbon farmers were spilling out of the inn onto the street.

Leena started to panic. "Chief or Yolanda, let's fix this thing now!" However, it was Bas-Mons who jumped down and went to work on the engine. In seconds, the parts had been pulled, flicked and reset so that the hover-cart could make its way back up the road to the biotower.

Hames was impressed. "You know your way around an engine?"

"Sure, been playing with them most of my life. We have a few shuttles to take us around the place, over to the other side of the valleys to the second sun-ray station. Sometimes to the outer zones of the citadel." Bas-Mons turned the cart sharply up another street and clipped a low post, sending small wood chips flying off the corner of the cart. "Sorry!"

"What do you mean the outer zones?" asked Leena.

"They're places where the coverage of the biotower starts to dissipate. Some of the carbon farmers still live out there."

Hames thought this sounded strange. Over the sound of the cart he called out, "Your dad makes a big deal about New Galapagos moving on and growing past its dome city roots. It seems strange to not having everyone living under the protection of the biotower and the sun-ray."

"Well, some people don't want to live directly under the Mons Administration. I've learnt that on the side of the street a few times." Bas rubbed his jaw. "In any case, they're not domes nowadays; they live in cubes, big glass cubes in the carbon mist."

Hames instantly recalled the image in his mind that he had seen and felt when he touched Marak Chakla – the image of a square in the clouds. "Did you just say cubes in the mist?"

"Yes," replied Bas. The cart took another hard turn, forcing the crew to hang on tight as it raced up sandstone streets lined with houses.

Hames spoke again to Bas, this time as if he were one of the Mons Administrators, talking clearly yet quietly. "Can you take me there?"

"Chief Hames, you're an interesting man, strange but interesting. Sure why not? I know the way."

The evening was completely upon them. The air lightly pushed against their faces and cloths as they approached the bio-tower. Bas-Mons had given Hames something to follow. He had no real explanation for this; it was more a perception. Regardless of the truth, he now held a strong sense of lucidity.

Chapter 22

Recruitment and Selection

THE MORNING AFTER THE ADVENTURE at the Original Tortoise, Hames walked slowly down the stone pathway that wrapped itself around the inner hollow of the biotower, hoping it would lead him to where breakfast was warm and waiting. The curving trail was as disconcerting as the discomfort lying in the bottom of his stomach. This feeling could be due to either the absence of food or the bear-hug he'd received the night before; deep down, however, Hames had a sense that the unexpected was occurring.

Marak Chakla had shown him an image of a square in clouds. It had seemed genuine, like the feeling of warmth he felt weeks ago with the light alignment drive accident. Hames had wondered if the experience was just his imagination. It had made no sense until Bas mentioned there were, in fact, "cubes in the mist". Buildings that served as living structures in the outer zones for the recalcitrant inhabitants of New Galapagos. All this made him believe that if his sensation from Marak Chakla was real, maybe so was Emmalyne's. It was the stuff of fantasy that Hames typically and regularly discounted. As if to punctuate his internal pondering of recent events, he came to find both Emmalyne and Olisa in his path.

"Good morning, Chief Hames," said Olisa. "Emmalyne and I are about to make our way to the west balcony to meditate, do you care to join us?"

Hames wondered if this was gloating. Healers were not meant to pander to cheap emotions. It felt like gloating. Emmalyne stood behind Olisa's left shoulder and did not offer a word. She looked tired yet relaxed.

"How are you?" Hames asked tentatively.

"I'm fine, Chief. We have to keep moving, we've a morning session to attend," said Emmalyne with the smallest of head movements, barely looking interested.

Olisa smiled. "Yes. We must keep going; others may be waiting for us. Have a fine day, Chief." A lump in his throat inhibited him from saying anything as the two walked away.

His connection and friendship with Emmalyne felt broken, as if sheared by a laser cutter from the mines of Lugus. It appeared to him that the distraction and anguish that Emmalyne had been experiencing was gone. Replaced with a detached calm. Also gone was the light and warmth. The old Emmalyne appeared to have been fixed, despite not being broken. It was true she had been struggling; the breakdown of Emmalyne had been replaced with the breakdown of their relationship.

A short distance later, Hames entered a large room and spoke to the porter at the door. "Some brown bread, navachado and mana beans please, and a hot chocolate." This order rolled off his tongue like it was nearly forty years ago. His favourite recovery food. In truth, he was feeling strong compared to some of the others around him. Francois sat pale and still, the large man seemingly concentrating on breathing. Yoland was worse; she was flailing and wilting. Her arms would fly in opposite directions as she flung her shoulders down and groaned. Robero looked amazing. Hames said as such he sat down at the table. "Robero, you look a different person today!"

"Finally, finally I think I have turned the corner, Chief!" Robero was quietly pumping both his fists on the table, his Italian accent accentuated in his reinvigoration.

"No, Robero, please don't!" cried Yolanda.

"I barely touched the table," said the newly fit Robero.

It was too late. Yolanda was up and moving in what looked to be a dash for a destination she may not arrive at in time.

The COO had joined them for breakfast and surveyed her crew with a shake of her deep brown hair. "I think I will tell Tar-Mons we need a rest day today. We can go see the Crystal River tomorrow."

Francois said nothing, providing only a barely perceptible nod which indicated agreement with the proposition.

Robero felt strong. "Just as I feel like a real person again you all become ill! Ah, such is life. Chief, would you like to come with me to see some of the villages and stores just outside the tower? The ones we saw yesterday from the carts on the way to the Inn?"

Hames felt bad. On any other day he would have loved to spend time with Robero visiting these places. Today, he had his own exploring to do.

"I'm sorry, Robero. I've got to make my way up to the ship and check on my repairs."

The COO looked at him with a quizzical expression. "Everything okay, Chief?"

"Fine, fine. Dao is making good progress and if we can get the repair engine purged this morning we can set things up so it might be workable in the next day or so."

"Good work, Hames," said the COO as she turned to Robero. "Best to let him get to it, we've got a lot to work through."

"Okay, enjoy the *Medens*, Chief." Robero was clearly disappointed and the hint of sarcasm did not go unnoticed nor was it unwelcome. It gave Hames freedom from guilt to move up to the *Medens* and, after that, on to the plan to find the cubes in the mist. These were somewhere in the outer zone and Bas-Mons had agreed to show him this place. Hames also thought there would be another benefit in travelling with the capable young man. He caught the COO's attention. "I think I may have found someone who can work on the *Medens*. Someone who can work with the repair engine."

The COO's face was tired but curious. Her deep-olive skin looked more worn than usual under the pressure of getting the *Medens* back on track for Mermai. "Sure, who is it?"

"I'd rather not say; not until I speak to them first."

"It's your team, Chief. A team that's short two people so do your best to get at least one back in there."

A short while later, Hames found Bas-Mons preparing a shuttle at the top of the biotower. Sitting at the base of the craft to the far point of the wide pentagon platform, the young man showed no ill effects from the night before. "Hey there, Chief, how was breakfast?"

Hames felt an unexpected gust of wind blow over him, flicking his loose white tunic and making him unsteady for a short moment. "Not too bad thanks. What you up to here?"

"I've got to head out to the other sun-ray station. Just to do some checks and modifications. Regular maintenance, you know how it is."

Hames saw his opportunity. "You must have done this kind of work for some time. Are you interested in seeing something else? Would you like to come up to the *Medens* and give me a quick hand with the repair engine?"

Bas-Mons dropped his normal carefree smile, as if he wanted to be sure this was true. He rapidly wiped his hands clean with a nearby cloth. "Absolutely. Yes."

"I'll tell you what, if you can delay your maintenance shift until later you can come up now. Maybe, in return, afterwards you could show me the cube living spheres in the outer zone. How does that sound?"

Bas looked around, possibly checking if someone was listening. "That sounds like a plan, Chief. Give me a moment and I'll be right with you." With excited movements Bas' long sinewy arms packed up his tools, placing them to a far side of the wide open platform.

A warning noise signalled Temett's arrival with the *Medens* shuttle, ready to swap over with Leena, who came up behind them. She slapped a brown hand on Hames' back. "My turn up top for a while, Chief. I'll take you and the boy back up."

Bas smirked at Leena. "If you haven't noticed I'm quite a bit taller than you. Not so much a boy."

Leena gave Bas a serious look as she stepped into the shuttle to relieve Temett. "With that tool muck on your hands you smell like a boy, that's all I need to know."

Bas-Mons gave her a big smile and winked. "Hey, I can't help it! New Galapagos is pretty humid you know."

After docking with the *Medens*, Hames took Bas-Mons into the repair engine room to work with the interface block, the recent focus point of disaster with the overflow of the particle conversion debris. The memory of the black and hazy material flowing into the repair engine room was still vivid in Hames' mind.

Dao Mai was there in voice, not in physical presence, as he was working on the outside of the ship. "I am soon to be finished with the exterior repairs, Chief Hames. In eighty-eight minutes I will be returning to the *Medens*."

"Thanks, Dao, I think I know what's going on now. Much appreciated," said Hames before turning his attention back to his guest. "Okay, let me show you around the place while I make a few fixes to the interface block." Hames spent the next hour with Bas-Mons taking him through the workings of the repair engine; work that would normally have been completed by Xavi or Chung. The pain at their loss crept up on him, but he slowly and methodically beat these feelings back down.

There was now an opportunity with Bas to recover the lost skills and resources the Medens needed to continue its journey. Bas was smart, incredibly smart. He understood the mechanics of the ship just as well as anyone, even though he had never worked with particle conversion technology before. They then spent an hour in the Repair Control Unit looking over all the data from the accident and what was left ahead of the Medens in repairing the Mermai Tentacle.

"It's amazing, what you are doing. Fixing space to see if we can get back to the colonies. A bit futile, though; it's going to take some time and you'd need safer particle conversion drive tech."

"It's worthwhile. We learn new things all the time and have some amazing experiences along the way." Hames was working up to his recruitment pitch. "Did you know that a month ago we encountered Ugor Eels?"

"No! Really? I wondered if they were just hoaxes. Did you see them?"

"Yes, they're real. The crew were pretty taken with them I have to say."

"I'd be taken too. A special experience for you all."

"We might see them again; they seem to be attracted to the Medens. If you came along with us you could have that chance."

"Ah. That would be nice but I could never leave my father. Since my mother died a few years ago we're all we have." Bas was wistful, and then his big smile returned to his tanned face. "I'm also destined to lead the colony of New Galapagos as lead administrator!" He sat down very formally with his arms wide and head held high, almost grimacing at this outcome.

"I had to ask," replied Hames. "I think we're done here. Let's go get Leena to take us back down to the surface and you can show me over to that outer colony."

"Sure, and thanks for this, it's been a real treat." Bas-Mons shook Hames' hand as if he were saying "thank you" for the visit to the Medens and also "no thank you" to the offer of passage.

As they walked through the corridors of the *Medens*, Hames attempted to assuage his curiosity. "I still don't get it. Why are there people living in biospheres in the outer zones?"

Bas was in a happy mood and willing to share. "Like my father suggested, we had an influx of people who didn't really support Universal Direction." He looked cheekily at Hames. "Yourself included. Some people from these contrary groups were encouraged to move away from the main citadel, to ensure their effect on the others was muted. Eventually they shifted to the outer zone away from the biotower where some of the old domes were still in good working order. Over time they reshaped them to be more efficient; a square shape was their choice of architecture, or a cube I guess."

Hames rubbed his shoulders. It had been a while since he'd undertaken so much physical manipulation of the *Meden*'s systems. "When I was here decades ago the biotower was finished and the sun-ray in operation. I don't remember any domes or cubes. Everyone was so happy to be rid of them."

"New Galapagos is a very rocky place. There are places where you can see for much distance. There are places you can't see at all."

"Why would people want to live in biospheres when New Galapagos seems to be thriving? And why does your father allow it?"

"My father is entirely committed to Universal Direction and a harmonious society. One theory for such harmony suggests that any secessionist elements need to be placated or possibly removed in order for that society to flourish."

It seemed an intolerant approach to Hames, although he assumed there were a few sides to the story to be told. They arrived at the bridge. Leena was checking the orbiting controls for the auto-pilot. She was surprised to see them. "You were quick!"

"Bas picks things up pretty quickly. He knows his way around tech that's for sure. I reckon just a few more synchronisation cycles and the repair engine will be back online and pushing energy across the ship evenly."

Leena took the news less positively then expected, rubbing her sentir spot as she spoke. "Right. I guess that means we'll push on to Mermai. Still, a few checks are required. I've been noticing power fluctuations are still knocking the LAD out; there's no way we could move into top speed right now."

"Like I said, a few more cycles over the coming days and we should be good to leave New Galapagos," responded Hames.

Leena gestured to the exit point. "Okay, you ready to head back down?"

Bas-Mons took on the role of guide for the afternoon's adventure. "Yes, but could you drop us back on the other side of town? There's a transport platform behind the Original Tortoise that the carbon farmers use. I thought we should apologise to the publican. We left there yesterday with, ar, unfinished business," Bas-Mons trailed off, looking clearly guilty.

Leena led them away from the bridge. "You did make a mess of that situation, Chief. An apology sounds about right to me."

Bas gave directions to Leena and the first pilot guided them down to the landing platform they had seen from the window of the Original Tortoise. The area was almost as big as the *Medens* and was sporadically populated by small oval-shaped clunky craft taking on and off.

"Whereabouts, Bas?" Leena asked, holding her hands up. "It looks like you can land anywhere."

"Just near that building over there." Bas pointed to the back corner. "The one next to the exit point to the main street."

Leena put the shuttle down next to a single-storey rectangular building that was stained with green and blue grime along its rusty brown walls. Hames and Bas disembarked and with a wave Leena left them for their undisclosed journey to the outer zone. They entered the murky building that also served as the exit point for the carbon farmers making their way home from a long day's work. Some of the farmers were making a direct exit onto the street, most likely to head up the hill to the Tortoise for a drink, still garbed in their farming attire. Others hung around, removing the bulky suits, sitting in their sweaty underclothes.

Bas and Hames found their way to a back corridor of the building. Bas spoke quietly. "You might have guessed the village we're heading to is fairly separate to the main citadel, so I'm not really welcome there."

"What do you mean? You said you'd been there before." Concern entered Hames' voice.

"I've been there, it's true. My curiosity couldn't be helped. I just couldn't go in as me now could I?" Bas-Mons' ever present happy smile was beaming.

Hames looked around and saw some carbon farmers' suits hanging up in a back area change-room. He now realised why Bas wanted to come here; it wasn't to apologise to the Original Tortoise publican. "Great, we're wearing the suits."

"Yep. They smell terrible."

In a large changing area, some of the workers were hanging their suits up and heading to the next room to scrape off the coating of grime left over from the day. "Quick, let's grab the two over there." Bas-Mons ushered Hames over to a far wall and after pulling the red bulky protection suits over their own clothes they both grabbed a helmet.

The suit was drenched in sweat and smelt accordingly. "Urgh" was all Hames could say. He felt guilty at taking the worker's gear, but also determined to see the village that was made up of "squares in the clouds". Robed in their newly acquired attire they walked out onto the main street carrying helmets under their arms. Hames felt energised despite the heavy suit. The idea of a group of ostracised dissidents living out in square containment pods fascinated him. To Hames, it was like a window back in time to before Universal Direction had taken hold. It was also a window he needed to look through in order to ease his curiosity. "Your father mentioned that thirty years ago some of these people set bombs off in a healing centre?"

"That's the story. Before my time, of course."

Hames side-stepped a street vendor's cart selling bug-wheat, nearly falling onto the sand-coloured cobblestones in the process. "Don't you worry about these groups? Maybe they could start up again."

"Father doesn't like to talk about the details too much. Most of the ring leaders were killed in a fight at the time of the blasts, the rest were sent away. It apparently was an agreeable outcome. They haven't been a problem for years. They still do some work in the community, like carbon farming, and we trade with them. So no problems, Chief, nothing exciting here."

As they stepped through a crowd of residents coming in and out of a grocery, Hames ventured one last try at securing Bas-Mons' time. "You know, if you're looking for excitement you could try the *Medens*. It's been pretty eventful over the last few weeks. You could learn a lot working on the repair engine."

Bas stopped in the street to face Hames. "Chief, the trip would be good. I just think that over time you and I would struggle to integrate with each other."

Hames was taken aback by this statement. "How do you mean?"

"Well, all my life I've lived and operated with people around me committed to my beliefs. I'm starting to see that as much as I enjoy your company you would never agree with or understand my choices."

"I've had Universal Direction around me for years. I've not had a problem with anyone." Hames paused. "Well, that's not completely true. I have had the odd debate. It doesn't make the ship work any less efficiently."

"Do you try to see the other side?"

"I think I do. It's hard to give in to something you don't believe in, just to make people around you happy."

"What about the things you have seen? The Ugor Eels that have started appearing, even near your ship. New forms of life. Are they not an example that you don't understand the Universe as you thought it to be?"

"I'm not sure about that. The eels, while new to us, still seem to be composed of a fundamental property in our Universe, electric charge. I'd disagree that they come from some sort of god-source or akashic field."

"Can't you see, Chief, that through your touch of your friend Marak there is a linkage there? Perhaps even with all the non-functioning of his body and mind he is still present. That seems to me to be a really good case for all things being enminded."

Hames felt a lump in his throat as his aged and preconceived concrete calculations of the world started to seem weaker than he remembered. Unconsciously, this was his body betraying a small yet burgeoning disbelief. This was as close to an out-of-body experience Hames believed he could ever get.

Bas-Mons reached to a nearby shop basket full of purple pears. He pulled two out; one for Hames, the other himself. He started to peel the dark fruit as they strayed further down the street.

"Your friend Leena, she seems to take a different approach than the others."

"That's true, she sees the value in the individual. The rights of the individual are a core part of her beliefs I guess," said Hames.

"To me, she's more in line with the idea that consciousness can be present in many things." Bas tossed the remnants of his pear into a nearby garden. "Look, thanks for the offer, but no thanks, Chief. My place is here with my father." Bas-Mons was committed to Tar-Mons and the administration of New Galapagos. Hames could see there was no moving him.

The two of them turned into a small avenue that pointed away from the main street. The brown and bent tree houses appeared more infrequently with most decreasing in size and grandeur the longer they walked. They stopped at a pole that appeared to be a transport point. They were headed for the outskirts of the villages of New Galapagos and an exit from the citadel. In quick time a hover-cart arrived, more beaten and worn than their transport of the day before. When Hames climbed on he was certain he would get a splinter. Bas took a seat next to him and two carbon farmers, also garbed in their full suits with helmets in hands, joined them on the small vehicle.

The hover-cart began an automatic path, devoid of driver and human control. It worked hard carrying Hames and the other passengers through the streets and finally away from the residential areas. Almost immediately, the cart began a surge up a dark craggy hill. As they passed over the crest Hames hoped to see the cube-shaped biospheres ahead of them; instead there was a slow dipping valley with a minor crest on its other side. Now the hover-cart had left the defined lanes of the citadel it began to scan its surroundings. Intermittently, the cart shot a laser from its lowest point that spread an electric blue beam across the ground, licking and knowing every rock for forty metres around it. Whatever direction the vessel may take it would know when to lift higher and drop lower.

Hames didn't know where they were going and noticed that their descent into another rocky valley brought with it a dip in the light afforded to them by the sun-ray. It was still warm, but without the sun-ray's direct influence the light was lessened.

Bas gave him a pat on the shoulder as they progressed further into the valley tapped the helmet in Hames' lap. Behind him the carbon farmers were mirroring Bas-Mons in placing their helmets back onto their suits. Hames copied them, realising that they were departing the oxygen-rich protection of the biotower. They ventured further into the uncontained landscape and mists began to present themselves, ever thicker and denser.

Hames looked back towards the sun-ray and the biotower; he could see neither. In front were vapours of white and grey. Over the next crest Hames' anticipation was rewarded with a vision of the squares in the clouds he had seen and felt when he touched Marak Chakla. Laid out ahead was an extremely large cube-shaped biosphere that had translucent access ways off to smaller cubes around it. All the square-like structures shone brilliantly from the intense light that pushed out from within. A quick count told him nearly forty cubes of varying sizes were connected to the larger cube, which Hames thought would cover a small playing field. All these structures were illuminated and green in the night, swathed in the carbon-infused clouds of New Galapagos.

A feeling of elation swelled inside him. He had not felt this sense of wonder since a child, not since he took his first trip on a particle conversion drive ship to Cernunnos as an eight-year-old boy. He gripped tight to the side of the cart, sending shards of wood into the air. They descended further and further into the dense clouds of the rocky vale.

Chapter 23

Power Squared

HAMES AND BAS-MONS APPROACHED THE glowing square that was the entry point to the village in the mists. Clinging to the outer layer were many chelonoida, attracted to its brightness, their shells pressed hard against the wall. The front of the cube destabilised itself to let the hover-cart passengers enter. Bas was constantly looking at Hames to make sure he was okay. Hames was too excited to see what was within the larger cube to be fearful of being found out. The two carbon farmers went first, walking through a long passage way without notice. Hames and Bas made their way to follow but a small tone rang out, a message distinct enough to cause them to stop and look around. Ahead of them the passageway had sealed itself. A man appeared on a screen to the right of them. Even through the viewer they could see he was small and very pale.

Superiority rang out in his voice and he viewed them suspiciously. "I haven't seen you before."

"We're here to visit a friend from the fields. We've heard so much about your village we decided to come see it," replied Bas-Mons.

"We don't get many visitors out here. We don't visit you either." The man on the screen peered at them down his long nose. "And who is the friend that speaks about us so much?"

"Joseph Prana, he's as enamoured with the Administration as much as I am."

"Ah, yes, Prana, a good man, good man." The screen went silent as the gatekeeper at the other end considered their request. Finally, he responded, "You can come through, but remember we have your biosignatures and I've set the exit so you have to ask to be let out. I'll be expecting your call soon."

"That sounds just fine," said Bas. The passageway ahead of them cleared to allow entry to a large living area. It seemed an immense world to Hames. He guessed it was about forty metres high and eighty metres long and wide – not so much a cube.

Surrounding the walls were stone benches and exit points to the smaller cubes. Hames and Bas walked over to the nearest seat to their right and started to disrobe. Out ahead of them was a trim grass lawn dotted with statues and ornamental pedestals. People were coming and going from the cube exit points. The two visitors breathed in the cool air. It struck Hames that the vast structure was a bright and vibrant locality. Despite the ostricisation of this community, it had a feeling of life.

"This is the main open area, the communal meeting place. The other cubes are where people live and conduct their limited business," said Bas.

"So you came here before?"

"Yes, one day about a year ago, a farmer called Prana clocked my chin in the Tortoise. All uncalled for; he just didn't much like the idea of the son of Tar-Mons drinking near him."

"Ouch."

"Yep, it wasn't the best feeling. I got curious about why he would be so anti-Administration so I snuck in here not long after that, much as we just did now."

Hames looked around the large rectangular space. "I can't see that gatekeeper anywhere, he must be in a smaller cube or another area."

"Who knows? This place is pretty big. I can't tell what's down the other end from here." Bas-Mons was nearly out of his suit. "The gatekeeper, don't worry about him. People from this community don't mingle much with those back under the biotower. We're strangers to each other, apart from the odd interaction." Bas-Mons rubbed his chin ruefully. "There seems to be an assumption that if we leave them alone they'll do the same, which is about right."

Hames and Bas started to make their way around the border of the biosphere. In their stained and sweaty underclothes, they looked just like the other carbon farmers coming and going through the main cube. They reached the far left corner and found a group of people talking and interacting. "There's a real community here, isn't there?"

Bas looked around them. "They seem happy together."

Hames walked closer and he noticed that despite the differences in body shape and clothing of the people, they all wore a pure white thick piece of rope around their wrists. "Carbon farming doesn't buy you much on New Galapagos does it? Their clothes look tired and old."

"I'd only ever observed they wear more than what we need under the sun-ray."

In the centre of the biosphere, something caught Hames' eye; it was an open paved area in the shape of a square. Extending out from it were four rows of seats from each side of the square. It was a familiar image although he couldn't place it in his memory. Bas-Mons seemed to be losing interest. "So, Chief, you seen enough? I don't really want to hang around here too much longer. This lot can have quite a temper if they feel impinged on in any way."

Hames looked around at the villagers bustling in and out of the exits. Some had baskets of food while others carried boxes of tools and reading devices. It was hard to tell from inside the large living space what the smaller cubes outside served as. He started to peer more into the detail of the large parent cube. Something was powering this village and it wasn't coming from the New Galapagos citadel. Conduits were visible through the translucent cladding of the cube and in-laid into its top cornicing. Realising that the conduits must be for oxygen and power, Hames traced them back with his eyes until he spotted what he thought must be the source.

"Over there, that one." Hames pointed to an exit point not far away from a back corner of the main structure. "Let's have a look."

The two interlopers made their way to the exit point. They looked through at the cables and conduits leading down a corridor to a smaller cube. Unlike some of the other buildings, this structure was not glowing or translucent. They walked down the few metres to the point of entry, away from the sight of others in the main living space.

"I know we can't see in but I reckon it's the energy source for the entire village. Bas, what do you think of this place?" asked Hames.

"Hmm, I see what you mean." Bas looked over the structure and back out to the main living area a few times. "All the oxygen, the power, it's a lot to generate and this village is massive. I think this is where it's all coming from; a power cube as it were."

"Right," said Hames, "and how does that much power come from such a small installation?"

Bas moved a step towards the door of the power cube and then stopped. "I'm curious, but this isn't our business anymore."

Up until this point, Hames had been on an exciting adventure, like an explorer discovering a lost tribe in the jungle. He should be respectful of these people, respectful of the COO and the rest of the *Medens*, and not cause any more trouble. But it was difficult. He moved forward to the front wall of the power cube. The door didn't open.

"The power cube not opening suggests we're not meant to go in there," said Bas.

Hames stalled for time. "Pretty sound logic; you'd go well on the *Medens*."

"Okay, time to go, Chief; you've had the visit you asked for."

The chief mission officer sighed silently and dropped his shoulders. It was anti-climactic. To the side of the power cube wall there was a panel that looked to him like an old manual over-ride. "Isn't that unusual?" He pointed at the box with its wires and levers.

Hames could feel Bas-Mons' curiosity overpowering his caution. "I guess so." Bas' voice wavered as did his commitment to leaving. "It'd be good to see how it works."

"Yes, well, it never hurts to look." Hames gave him a cheeky grin. Bas-Mons flicked the panel open and began to play with the wires and parts till suddenly the cube entry wall was activated.

"Bas, I want to agree with you that this is wrong but something tells me we're going to enter the cube." Hames could not see all the way through; the cube was bigger than most of the smaller cubes but was filled with partitions of darkness.

Bas-mons breathed deeply and nervously. "A quick look and then we leave. I don't want to cause any problems, Chief. If my father found out what I was doing I would be in more trouble than I could quantify."

The serious look on his face convinced Hames this was true. "Sure. We'll have a quick look, that's all. Right?"

"Right."

They entered the power cube and walked around the first partition to find a control centre. "Well, that's expected, they have to manage the flow of energy somehow," said Hames

"Not very interesting if you ask me." Bas-Mons was also disappointed. They walked around another partition to find a row of cabinetry that held various tools and safety gear. A further search revealed only the usual mundane equipment needed for a generation source.

Bas started to relax somewhat. "So far pretty similar to what we have back in the citadel. I'm just waiting to see what is enabling this place; it can't be bio or closed cycle efficient, it's too small a location."

Hames was of the same mind. "I'm guessing that like on Earth, New Galapagos ran out of Lugurian minerals decades ago."

Bas confirmed the assumption. "We never used them in the first place."

As they moved onwards a familiar isochronic hum resonated inside Hames Naughton. This brought into him a feeling of unease. Hames passed the final corner of batteries and controls he felt his anxiety turn to fear and clarity at the same time. What was in front of him was unmistakable. He had studied them since he was a young man. Tried to build one when he was thirteen. Succeeded when he was sixteen. He had built, reconditioned and redesigned many since that first achievement. He knew all types. Not only how they worked but which stars in the distant galaxy they could take you to. He had been there when they failed. When they broke and destroyed the paths to friends far away. When their failure had taken his soul away. He had been there when they were outlawed like a criminal, chained and chastised. Hames' entire existence of the last few decades had been spent trying to fix the mess they had left behind. To his horror, powering the village in the outer zone was a particle conversion drive.

Hames had never seen one in operation at a terrestrial level. This alone provided enough fear to freeze him in place. His distress was escalated for other reasons, though; the particle conversion drive was made up of old and ill-fitting parts. It was half built into the back wall of the cube, exposing itself to the outer environment. In this sense it was ingenious as there was enough exposure to the ground atmosphere for it to convert the requisite matter to generate power.

"Chief, is that what I think it is?" asked Bas with trepidation.

"Yes. Yes, it is."

"Shouldn't this not work? Shouldn't it be causing a massive explosion?"

"Only if it tried to convert any of the matter broken in the Cataclysm. It's so small it's not reaching high enough outside New Galapagos to touch the Mermai Tentacle. It must be close. So close," Hames whispered, his anger growing.

"How long do you think it's been here?"

"I don't know. It's not going to be here much longer."

"What do you mean?"

"All these people are in danger. Terrible danger. The Mermai Tentacle surrounds the entire New Galapagos system, all the way past the fourth moon. This engine, this whole village is still here by luck!" Hames was pacing in his fury.

Bas-Mons was starting to follow Chief Hames Naughton's logic. "Okay, I can see they're four magnetic plates on the outside pulling in matter from the lower atmosphere to be converted." The young man ran over to the far side of the room to look at two output pipes. "Out here is the converted matter. This converted matter looks like it's being channelled towards the outer atmosphere. I guess it eventually joins the other parts of the tentacle. You're right; it's lucky they haven't blown themselves up. They've designed it, so maybe they're aware of the risk?"

Hames ran his hands down his sides as if he was about to start work. "Bas, there is no question about it. The risk is too great. This thing needs to be shut down!"

"Let's see if we can't check out the space around here; they could be converting such a small amount that they'll never touch the Mermai Tentacle. This is this village's life blood, we can't be too quick to act."

Bas and Hames rushed back past the partitions to the control area and started reading the space around them. The answer was not the one Bas was looking for. "It says they've had this system for over twenty years now. In the last few years they've come close to converting damaged debris from the tentacle multiple times; it really is pure luck this place is still here."

Hames began to enter in his own code to reconfigure the system. "It's got to go!"

"In good time, Chief. In good time. It's not your decision!"

Hames was making quick assessments of the particle ecosystem outside the cube village. "Just look at this: only a week ago they took matter within a metre of the tentacle. They nearly died then, all of them! There must be hundreds of people in here!"

"They must have safeguards! Look at how they dispose of the converted matter – it's kept well away from the conversion drive." Desperation was in Bas-Mons' voice as he argued with Hames over the future of the cube village residents.

Reluctantly, Hames checked the system in front of him for safeguards. He could find none. He then looked at how they disposed of the converted energy and his blood began to boil with a final realisation of the damage that had been done by this conversion drive. He began to shout. "This thing has been allowing post-Cataclysm debris to float up to the Mermai Tentacle for years, decades!"

"Why is that such a bad thing?"

"By pushing more converted particles into the sky it's fundamentally changed the damaged space in the hundreds of thousands of kilometres stretching back from New Galapagos to Earth."

The implications of this finally dawned on Bas. "Chief, think this through, please!"

Hames should have seen it straight away. This was the reason why the repair engine on the *Medens* had failed. Without thinking any more, without any more exasperation, he finalised his termination code. The rickety particle conversion engine stopped.

"What have you done? This wasn't your decision!" The pain in Bas-Mons' voice gave Hames a moment of concern that soon vanished in the knowledge that this power cube and its particle conversion engine had killed his friends.

"We've got to get out of here, Chief! These people are angry at the Administration for just being around, let alone switching off their power source!"

"There's a wall of batteries. I estimate they've a few days' power in there. Enough time for you to help them find a different source."

"They'll just start it up again, Chief, you've achieved nothing!"

"My termination code won't just switch off the machine. Right now the magnetic pulldowns are warping, the molecules of the internal parts are destabilising and all remnants of computer code are being wiped.

They would have to completely start again, from zero." On cue, Hames could hear the plates shifting and the loud smack of metal. Cracks ignited in parts of the machine and it began to seep a deep dark energy, similar to the one Hames had seen take over Xavi.

"Let's go, let's just go," implored Bas-Mons.

Hames relented and they made their way out of the power cube and through its front passageway to the main cube.

As soon as they entered the wide-open space a throng of angry villagers came towards them. They knocked Bas-Mons to the ground with a long wooden club. Hames then felt the air move behind him and a numbness drove through the back of his head, sending him away from the light.

Chapter 24

Hammersmith Harmony

Nisha surreptitiously shivered. As Hames' partner, she was positioned in the first row for the graduation ceremony with a direct view of her love, who sat on the podium. Unfortunately, it was a cold auditorium and her simple floral dress, while matching her deep mermain colours, did little to raise her body temperature. *Get on with it!* she thought as her teeth chattered and clenched.

Hundreds of family and friends were present to watch their loved ones graduate from the Agency Academy. Speeches and awards had flowed, as had much hand shaking. She twitched and pulled at her attire, frustrated that all this fanfare was both freezing her and eating into their packing time. Getting home now was more important than ever. Earth had become a tense place. Hames didn't see this. He had asked whether they could stay a few more weeks, to let him work on the new particle conversion routes from Earth. At this suggestion she had put her foot down, literally. Her wincing at the pain in her shoulder had reminded her partner why she was so keen to move on.

The bomb blast at St Paul's as well as her interactions with Universal Direction and Montaxe-Deus had shown her that Earth was like a simmering pot of water, soon to reach boiling point. Strangely, although she felt this keenly, the community around her did not. On her way to the auditorium in Hammersmith there was no such collective feeling of concern on the sky transport. Conversations had drifted from the ordinary to somewhat interesting topics. It seemed that Universal Direction had not fully infiltrated the thoughts of the local people. It was only after the short walk from the Hammersmith station that she started to hear more talk of Universal Direction. The students of the Agency sounded extremely familiar with the new approach to a harmonious society. Perhaps their cloistered position in the Agency provided them a deeper education of the new conventions. Hames had mentioned something about daily sessions being offered but he'd skipped these to continue his studies.

Since the ceremony began, Nisha had been dreading the final speaker, Legeles de Nord, the leader of Earth and Universal Direction. It was uncomfortable being reminded of the day at St Paul's. Eventually, he was introduced to applause. Legeles de Nord strode confidently towards the centre of the podium, pristine in a sleek long grey suit that aligned with his long grey hair, his arms raised to the side recognising the goodwill afforded him. Eventually, he quieted his welcome and stood in silence, watching the room in an almost personal moment. To Nisha's shivering relief he finally began.

"The seven colonies and Earth. Eight worlds. Eight communities in space. Connected we are. Not only through ships that ferry us back and forth, across time and space. We are all simply connected." Legeles de Nord pushed his fingers together to intertwine them in a very demonstrative show of intractable linking. His voice grew in strength and gravitas. "Earth is as much of Mermai as Mermai is of Earth. Cernunnos is as much of Lugus as Lugus is of Cernunnos. We on Earth must show the way to our brothers and sisters removed from this place. We must show them they are part of us and we are part of them. We are everything. We will never forget them."

Much clapping ensued at this point. Nisha felt this was misplaced as the last words from the speaker sounded more like a threat than a promise of hope.

"We have identified the method and path we can take. It is one that has always been there in front of us. Harmony. Harmony in our community. Harmony in our life and, most importantly, harmony in oneself. If we ourselves cannot be harmonious within, well..." Legeles de Nord paused and gave a wry smile. "We cannot live with others in peace." More clapping followed and quiet murmurs of agreement. "Universal Direction has been provided to achieve this goal and already I can feel its effects. Collectively, there is a beautiful feeling that things are changing. Individuals who once were harmful now are helpful. Whole communities of the disaffected and deprived now feel safe and satisfied. Harmony is happening."

More cheers and cries made Nisha realise she had been following a different media stream in her life to the ones of the students of the Agency. Hames too was oblivious to these happenings. She felt somewhat perturbed that he had not been following the progression of Uni-D amongst his classmates. It was good he had his own mind. At times, though, he needed to be more aware of what was going on around him.

"There are some who would try to halt the move to harmony. Some who cling to the old ways of dissidence and disquiet. We will not lie down, we will not accept a discord with the opportunities to turn anew the ways of old." Legeles de Nord looked to one side and shouted out the first convention of Universal Direction. "The past is past!" to more applause. He then swung 180 degrees to the other side to the same effect. "We are here and now!" Clapping was now raucous and Nisha felt nervous rather than elated as the speaker continued to rile his crowd. "What will be, will be!" The applause reached a frenzy as Legeles de Nord pointed to everyone present. "You are the everything!"

Please just stop, Nisha thought to herself. More minutes of adulation passed and then at last the proceedings were over. The class of engineers began waving to the crowd and bowing as they were presented to the audience. Thirty-two in all looked out from the stage to the large auditorium that beamed daylight through from the full glass ceiling. Hames was standing in the middle as the valedictorian; he seemed content and looked only at her. With this acknowledgement her happiness and relief started to return. "No tears," she said as she bit her bottom lip.

It felt like the beginning of the end of their time on Earth. She had waited a year for this moment, maybe even longer. Nisha started to clap even louder, just as a few people stopped and started to sit down. She continued while more people sat down. A short moment later she realised she was the only one still standing. Her clapping slowed as she looked up at Hames. He was beaming at her with what she knew was unconditional love. It must be, because she was most likely embarrassing him. Yet there was no avoidance of recognition, just affection, and she happily took her seat. The master of ceremonies then closed the event to loud cheers and all involved arose, exiting to an open paved area and garden nearby.

By the time Nisha entered the concourse outside the auditorium it was densely packed with the friends and family members of the

263

graduating students. Without much room to move she made her way towards the side of the riverbank to look out and down the Thames. Afternoon sun beamed down onto Hammersmith Bridge to her right, catching and warming her exposed shoulders. The clear blue water beneath was flowing fast and fluidly, almost lapping at the top of the pavement. A slight wind came off the water and the feeling of being in nature seemed to remove Nisha from her recent worries over Jasiah, Montaxe-Deus and Universal Direction. She was over all of that. It wasn't her responsibility and she still couldn't fathom how she had become so involved with it all.

She turned around to look for the students and spotted an older woman wandering in and out of the concourse crowd, saying hello to various people. The woman looked familiar. She was wearing a red and white striped skirt, her grey hair was pulled back loosely and her blue eyes locked on to Nisha. The woman moved with great agility for her bent body. She extended a hand to Nisha as she reached her. "You are Nisha. I have seen your face many times before." Her voice was cracked yet filled with calm confidence.

Nisha was unprepared for this and felt almost flattered. "Oh, have you? That is reasonably uncomfortable to hear."

"I am sorry to embarrass you, my dear. I have spent many hours at the desk of Hames Naughton going over his calculations and theories for particle conversion routes, they are marvellous indeed."

Something clicked in Nisha's mind. "I'm sorry, I have seen your face as well. You're Bala Garang, one of Hames' instructors."

The woman identified herself to Nisha with a small bow of her head. "Yes, that is right."

Nisha contained a smile. "Just to reassure you and myself, I haven't seen any photos of you in my living-pod."

The older woman laughed. "I would hope not."

"No, but I have seen you in many vids. You were the first person to map a PCD route, to Earth-2 and the route to Mermai! You're absolutely famous on my home planet! The pictures of you with the cartography scans of Mermai were in my family's hallway. I didn't recognise you, you..."

Bala Garang cut off Nisha's sentence. "Have gotten a lot older since then. Yes, that was over a hundred years ago. I do look old. It happens to all of us. No matter how well we care for ourselves."

As she had done many times over the past weeks Nisha pulled at her necklace, running her fingers over the shape of the mermaid Sovann Macha.

"That's a nice pendant; a pretty mermaid indeed."

Nisha held out the leather strap and pendant to show her. "It's Sovann Macha, she's from an old Hindi fairy tale."

Bala Garang peered inwards at Nisha's pendant. "Hmm, I'd heard these symbols were adopted quite religiously on Mermai not long after it was settled. She looks very important and pretty with all that colour."

"Thank you, she is very important, at least to me."

"You know, as I recall it, all the heroines in old fairy tales have one of two endings. They either get the prince and live happily ever after, or they end up lost and alone. Confined to some old tower or lake, their spirits enchanted and tied to their holdings."

Nisha liked this old woman; she was smart and funny. "I guess so. Sovann Macha's is a story of love. She was a daughter of the demon Ravana, who stole Sita, the wife of the God Rama. Ravana kept Sita on the island of Lanka and to save his wife, Rama asked Hanuman, another god, to help rescue Sita. Hanuman and his monkey warriors started to build a bridge to the island of Lanka by throwing great boulders into the sea. But no matter how large they were, the boulders kept vanishing. Sovann Macha was in fact drawing them away so no bridge could be built. However, when Sovann Macha met Hanuman and heard his mission she fell in love and allowed the bridge to be completed."

"That is a beautiful story. Did they live happily ever after?"

"Ah, no they didn't. It's a bit complicated. They had love for each other, and they made a baby together but no, they didn't live together forever."

"See, I told you my dear. Fairy tale queens, lost and alone. Ghosts to what could've been."

Nisha was unperturbed. "I keep the pendant because she reminds me of home and how a bridge was built from Earth to Mermai. There's also the fact that the main continent on Mermai looks like a mermaid." Nisha was laughing at how silly she thought she sounded. "You know all about that, though, being the person who planned the route there!"

Bala Garang was laughing too. "Ah, I've lived off the glory of those discoveries for some time. It's nice that they're not forgotten." She patted Nisha's good arm in a warm manner, much like an aunt or mother would do.

Nisha paused her discussion to look around the crowd for Hames. After a moment of silence, Nisha realised she had no more insights to offer the person who essentially discovered her home planet. Just absent-minded small talk. "The transport routes above have begun to slow; I think I may take the tube home today."

Bala Garang agreed. "Yes, it has been very busy lately. My man was telling me about it the other day. Said he was stuck on a sky transport for sixteen minutes. Couldn't figure out why it didn't just fly around the delay. He was so bothered by it all he stopped off at a healer on the way home to ease his anxiety."

Nisha was shocked. "So you're moving towards Universal Direction?"

"Young woman, at my age the thought of passing to dust and nothing more is very challenging. I much prefer the idea that I will not cease, rather I will be part of everything that is now, in the past and the future."

Metres away from her, Nisha saw Hames come through in the last group of students. He was immediately mobbed by all his classmates; his hair ruffled and arms all around him. He looked over at her and waved; however, by this stage he was being dragged away. Hames motioned his hand in a cup shape, a signal that suggested he would be going for a celebratory drink and she mouthed back to him "okay, love you" just as he was pulled vigorously out of the concourse and away from sight.

Nisha felt flat at Hames' departure. He could have at least said hello in person. Next to her, Bala Garang also looked despondent.

"Nisha darling, now that your man has so rudely left you for his colleagues, I will have to ask you instead."

"Ask me instead?" At first confused, Nisha realised she was more prepared for different these days. "Oh sure, please go ahead."

"We are quite keen for Hames to start work as soon as possible. I know it is short notice, but it's been four years since the Eagleton Accord and we do need to establish new destination points. Is there any chance you could depart for Mermai in the next two days?"

There was meant to be a week left before their departure, but to cut this short would be perfect. She did her best to contain a small smug feeling that pushed into her face and voice. "Bala, that will not be a problem. I will go home and pack today."

"My staff can reset your arrangements. I gather you are booked on the *Python*; it makes multiple trips each day to Mermai so it shouldn't be a problem changing your tickets."

Nisha waved her hand. "Oh, that's fine. I'm a coordinator at the Dover Shuttle Docks. I plan trips for people on the PCD ships all the time."

"Ah, that is fortuitous. The *Python* will be carrying much of the research load for Hames' examination of further planets past Mermai. I was going to give him this data chip." The older woman pulled from her pocket a small flat blue square with a metallic edge. "This holds my latest research on the travel coordinates for all current and potential routes to outer systems as mapped from Earth-2 and this Earth. Please can you make sure Hames gets this?" Bala Garang handed the square to Nisha.

"Thank you, I will. And thank you for your time with Hames. I know he greatly appreciated your advice." With no pockets in her dress, Nisha placed the small blue chip under the back part of her black communicator bracelet.

"You are welcome. He is a fantastic mind. Not since we first started mapping routes to other worlds have I been this hopeful that more can be found. He is the one to do it." The old woman wagged her finger at this last observation and walked away through the remaining crowd of students and their families.

Nisha made her own departure with a feeling of relief. It was definitely time to return to her and Hames' living-pod and start packing. She found her way to Hammersmith station and reached to the tube platform, ready to catch the District line towards Victoria. As she boarded the train she was joined at the last moment by Sean Watson, the man from the Agency who had convinced her to follow Jasiah to the Montaxe-Deus meeting in Mordon. It wasn't until he sat down next to her that Nisha realised who it was. She quickly turned away. "Enough of this."

The tube doors beeped long and high, closing as the train commenced its journey.

Sean Watson looked unusually disheveled; he was unshaven and his suit was somewhat dusty. "I just wanted to say thank you for checking up on Jasiah."

She turned back to him and snapped, "I didn't do it for you! I did it to see what Jasiah was up to. Do you know they scanned me?"

"That was expected, there was only a slim chance you might get by unnoticed." The now messy looking man from the Agency looked back at her apologetically.

"It could have been dead dangerous for me. I'm lucky it wasn't. Just unlucky to find Jasiah wasn't in a great mood to talk."

"He's sticking with them, isn't he?" asked Sean Watson.

Nisha thought back to Jasiah's direction to her to go home and to leave him. "I guess so. He asked me to leave him alone, so I'm doing that."

"Yes, I think that's the best option." The Agency man sounded drained, his voice matching his appearance.

The train pushed through the tunnel, passing pipes and corridors. Some of the angry red passed out of Nisha's deep-olive–skinned face. "Why are you here, Mr Watson?"

"There was an attack on a healing centre, the one you went to in Battersea."

"No! What happened?"

"We're not completely sure. It looks as if it came from one of the patients who visits there, Christopher Harkness."

"I met him, the day I went there!"

"I wanted you to know in case Hames hears and mentions it to you. Healer Milar was badly beaten."

Nisha was in shock. "Poor Healer Milar! He was so trusting. So committed."

"I've just left there, it's an awful mess. A lot to clear up. I wanted you to be aware that no-one is watching you now, Ms Pinto. We've been reallocated to protect the initiatives under Universal Direction. You're free from our protection."

"Thank you. Thank you for telling me." The train pulled into Victoria station and Nisha departed, grabbing the side of the carriage as she alighted. The news about Healer Milar and the urging of Bala Garang for Hames to leave early had filled her with adrenaline. She jogged her way through Victoria station and after a change of trains, found herself walking home from Battersea Park. They were now leaving for Mermai in two days and the packing would have to escalate. She arrived at her apartment and entered the passageway down to her living-pod. The cold, stormy long-lasting night of the Quadrant priesthood era came down around her. Broken walls and pipes filled her limited view and she struggled to find her door in the damp dank darkness. The themed corridors of her building were one thing she would not miss.

Nisha closed the door behind her and immediately changed into her plain grey pants and white singlet, bracing for an afternoon of logistics. It wasn't until many crates later that her stomach began to tell her of the need to eat. She sent Hames a message on her communicator:

> *"Hi there – sorry to bother you, my love. I wanted to check if you wanted any dinner tonight?"*

She waited a moment and the response was surprisingly quick in coming back.

> *"Hey, no dinner needed for me, having a liquid one here! Did you need me to come back?"*

Nisha responded quickly.

> *"No, miss you loads but it's our last two days now, your instructor has moved up the departure date so I'll have you to myself soon. Go, enjoy!"*

> *"Righto, love you."*

With the the final response from Hames it was time to look after herself again. A pantry of Mermain meat and chocolate was not going to be a very nutritious supper, but it would have to do. She looked at her communicator bracelet on her right wrist and considered ordering in. She began to call the local Cernunnosian take-out but was interrupted by the tone of someone downstairs at the bottom of the apartment complex.

She turned on the screen next to her door to see a teary and tired-looking Jasiah. His face lit up at her response.

"Thank God it's you!" he screamed at her.

"Of course it's me. I do live here you know!"

"Are you leaving tonight?"

"What?"

"Are you leaving tonight to go back to Mermai?"

Nisha was still feeling bruised from the horrible way Jasiah had treated her in Mordon the other week. "Why would it matter? You didn't want to speak to me again!"

"Please tell me, are you leaving tonight?"

"No, you tell me what's going on, then I'll tell you."

"Something bad's happened." Jasiah's head sank on the other end of the communicator. "I helped get them access to your profile, told them what you do. They're going to make a statement with a ship back to Mermai. I panicked when I couldn't remember if you were on it or not!"

"You're not making any sense. Come up and talk it through with me."

A short time later the buzzer of Nisha's door sounded, allowing Jasiah entry. She could not miss how dirty his white shorts and t-shirt were. The normally clean and curled brown hair was clumped while his green olive skin seemed to be fading. "Cousin, you look like a right mess."

Jasiah was out of breath and sat down on Nisha's bed. "I didn't think anything bad would happen, that they're all just talk. But after what they did to the healer's centre in Battersea I knew they were real. I knew that I had to tell you!"

Nisha's confusion was written all over her face. "Do you mean to say they attacked Healer Milar? I was told it was one of his patients."

"I'm not sure who it was, just that a few in our group started to claim it was them."

"Why would you think Montaxe-Deus are going to hurt me?"

"Not you personally. Remember, though, they have your bioscan from the meeting the other week."

Nisha started to follow what was going on. "They can't replicate it; in any case, the scanners would pick up a fake."

"I don't really know how they're going to do it. But when I told them about your work at the shuttle docks they saw an opportunity for access. To destroy a particle conversion ship. I really did think it was all talk."

Jasiah buried his head in his hands. "This afternoon things started to look real with people talking about attacking another healing centre."

Nisha stepped around her small square living space, her hands working underneath the back of her white singlet as she paced. "But which ship, and how?"

"They're going to take over your biosignature and use it to get into the PCD hub up in orbit."

"Jasiah, you're not making any sense."

Nisha's cousin began waving his arms. "One of them laughed; they said 'a Mermain attacking a Mermain ship' and the other said 'cutting the head off a snake'. That's all I could hear. I asked some questions but left before they became suspicious of me."

Nisha stopped pacing. "The *Python*, that's the only thing it could be. All we have to do is contact the shuttle docks. I'll go there myself if I have to."

"No! You need to stay away, that's the reason I came here! After the healing centre do you think Montaxe-Deus would stop at hurting expats from the colonies like you and me?"

"We can't stand by and let this happen, Jas; it's people's lives!"

"They will know I told you, and, they're using your profile to get in! You want to be somewhere far away from the docks to prove you've nothing to do with anything that happens there."

"Bugger that! We've got to sort this out!" Before Nisha could switch her communication bracelet online Jasiah was running out the door of her living-pod. She grabbed her navy blue work jacket and rushed after him. As she entered the corridor she could see him running away down the damp corridor, shouting "I'm sorry" over and over again as water splashed around his thick black boots.

Nisha locked her door and began to sprint after him, pulling her jacket on as she went. Moving awkwardly while she ran Nisha slipped in the wet cold water that served to enhance the corridor theme this week. Her feet flew forward, taking her horizontally into the air before she returned to the ground with a dull thud. Her shoulder began to sting intensely as images of the nefarious Quadrant priests spun around her. The pain surged and then dissipated as she sank into her own unconscious world.

Chapter 25

Cube Conflict

PAIN THROBBED IN THE BACK OF Hames' head. A sense of deja vu washed over him. It had only been weeks since he had been thrown around in the turret viewing room of the light alignment drive. He remembered that incident as a different form of unconsciousness, one that was cloudy and warm. By comparison, the after effects of being struck on the head just left a ringing pain, restricting his eye movements and seizing up his mind. Nowhere on his body could he feel the cool recirculated air of the main biosphere. He didn't know where he was.

Finally, he summoned enough energy to open his eyes and look around. The room he lay in had a minimal source of illumination; the soft green glow helped identify the shapes around him. Lying on a set of woven mats next to him was Bas-Mons. Both were bound with their hands behind their backs. The curled and long brown hair of the son of Tar-Mons fell around his face, obscuring his condition."Bas," whispered Hames. "Bas!"Slowly Bas-Mons groaned before moving his head to look groggily at his older greying colleague. "Chief, you look terrible. Urg, what happened to your eye?"

"I think a club similar to the one that cleaned your clock caught the side of my head."

"And look, your clothes. That pretty white suit of yours looks more like a Mermain cow with all the brown spots."

"I also ate some dirt when my face hit the ground." Hames rolled onto his belly and spat out some remnants of the village grounds from his mouth."

"You should be thankful that the past is the past and what will be will be."

"Don't quote Uni-D at me; not now, Bas."

"You're lucky I practise harmony, Chief; it's not what I am feeling for you right now." Bas-Mons was understandably angry at the events that had led to their predicament. "I think we're in another cube; it's hard to tell, it's pretty dark in here." Bas angled his head in different directions to get a better view. "The green glow is coming from what looks like a window frame. I reckon the power is still on outside."

"Not for much longer."

"Its nothing to be proud of Chief."

Hames lay on his back as best he could, looking up at the ceiling. "Bas, you're pretty committed to Uni-D but you're still, what, early twenties?"

"That's right."

"Have you had Curate-mente?" asked Hames.

"This isn't the time for these type of questions."

"Maybe, we're not going anywhere."

"You're still thinking about your friend Emmalyne, aren't you?"

"Yes."

"As your friends Yolanda and Francois said back in the Tortoise, it's a beautiful moment to see the Universe. It lets you see we're all connected so we should live in harmony."

"Perhaps. Maybe I'm the unharmonious one."

"That sounds about right to me." Bas-Mons scowled at Hames as he lay on the side of his head. "They're taking their time. I'm guessing they know what you did."

Hames coughed the last piece of dirt from his mouth, black granules catching in his beard. "That is not much of a guess. We're in some trouble."

"Chief, I think you mean you're in trouble." The young New Galapagoan sat up, his hands bound behind his back and his curled hair falling back to reveal his angry face. "If you had just waited for a while, we could have spoken to my father about this. Planned a solution."

Hames felt no remorse; his short grey beard bristled. "A planned solution? This group planned to blow themselves up! That's what was going to happen! Their makeshift engine perverted the entire space around this system. It caused the *Medens* to fail. It killed my friends and might have stopped the *Medens* mission entirely. Some of the crew want to turn the ship back around, Leena for one."

"She seems to be a strong and confident young woman. Leena wouldn't be afraid, would she?"

"It's not about being afraid for Leena. It's about people. She's wants to protect everyone. If she thinks the crew is in danger, she'll push for the mission to end."

"That sounds perfectly normal to me. If people are in danger, why go forward?"

"The Universe has been made small without access to the colonies. Leena doesn't understand this. I don't think she want's to be on this journey."

"Can you blame her? From what I can tell she's here by direction, not choice."

Hames considered this for a moment as he lay back on the woven mat. "Everything I've done over the last thirty years is tied up in this journey. Tied up with the *Medens*."

"What is the *Medens*?" A voice came from an older man who had entered the cube. His white spindly body extended halfway into their holding area, cautiously looking down to check that Bas and Hames were still bound and incapacitated. An increase in the light from the half-open entry point was enough to tell them it was the gatekeeper who hours ago had given them entry to the cube village.

Hames looked up. "The *Medens* is my ship, from Earth. We're here on New Galapagos on shore leave."

The small man looked down his long nose at Hames, his voice still as old and cold as it was on their first encounter. "What are you doing all the way out here, apart from damaging this village?"

"What are you doing out here damaging the Universe with your time bomb of an energy source!" shouted back Hames.

"You'll fix our engine if you know what's good for you!" shot back the pale-skinned gatekeeper, anger rising in his voice.

"It's not possible, you should look inside it. The parts are warped and destroyed. It can't be fixed," said Hames.

"Fix it!" the spindly man spat at Hames.

Bas-Mons sat up in his bindings. "Listen, it can't be done. He's telling the truth. If you let us go I'm sure we can plan a solution to your energy needs. We just need to talk to my father."

"Oh, we contacted your father an hour ago. He should be here any moment. We will not stand for this unwarranted and heinous intervention. There will be consequences." The gatekeeper pulled over a cane chair and sat down. "Why did you come here to our village?"

Hames struggled to explain how he came to be in this spot. He had no reason to lie so he choose the truth. "I saw an image. No, that's not correct. A man I used to know told me about this village. I came to see it. I just didn't expect to find a particle conversion engine tearing at the fabric of space."

The door behind Hames opened and in came a large balding man with a pug nose that looked battered from its life experience. His tan t-shirt was loose fitting, increasing his look of physical menace. Immediately, the new entrant to the cube saw Bas-Mons and struck him in the ear, knocking the young man flat to the ground.

"Prana!" shouted the gatekeeper. The burly man retreated to the wall behind the gatekeeper, leaving him to continue questioning Hames. "Did Tar-Mons send you? You have his son with you."

"No, they have nothing to do with this."

"Why did you come here? Tar-Mons promised to leave us alone."

Bas-Mons was coming around again to consciousness and tried to answer. "My father has kept his promise, he has! This man here is not part of our Administration. He acted alone."

The gatekeeper respondedly snidely, "Yet, you are here together. Are you not, young Mons?"

"That doesn't mean Hames Naughton is from the Administration."

"Your Administration, pah! The Mons family and its enforcement Universal Direction be damned." The white-skinned man shared a disapproving look with Joseph Prana. "You tried to turn us away from thousands of years of history. The past is important – we have to see the past to understand the future. After the conflict decades ago we were asked to move." The man emphasised the last few words. "So our cubes are outside the reach of the biotower and Universal Direction. We have done well for many years without the involvement of the Administration. Until now."

"I can assure you my father does not want to be involved with you either. He'll help fix this and leave you alone again."

Jospeh Prana again shaped up to hit Bas-Mons, but the spindly man held up his weathered pale hand to stop him. The gatekeeper then stared at Hames intently. "For years my family has looked after the particle engine which supplies the village. I admit I can only maintain it, not build one. So you must fix the mess you created!"

Hames was unrepentant; even if he could he wouldn't acquiesce to the request. "Not going to happen. It's over. The engine is broken."

The gatekeeper exited the cube, huffing and breathing heavily. The two captives were left to be overseen by Prana. Around ten passed until the old man returned. "Prana, bring them!" The large man grabbed both Hames and Bas behind their necks and dragged them out of the small dark cube and back out to the main biosphere. Hames again felt the coolness in the air and a gloom spreading across the now only partially illuminated village. Hames was responsible for this. He felt a hint of regret but quickly smothered it. The villagers' lives had been saved, whether they liked it or not.

Prana pushed them towards the square in the centre of the main structure. They both hit the ground awkwardly, landing onto their knees. Standing around the area were what Hames guessed was the population of the village. Men, women and children, all looking deeply disconsolate and angry. Nearly two hundred people stood all around them. Some crowded back towards the entrances of the other cubes. It was hard to tell if it was a show of force or if they were curious about what had happened to their home.

Hames looked to the far side of the structure and saw a group coming towards them, from the direction of the main gate. Even at a distance he could make out the forms of Siggy Conson and the COO. With her was Temett, Francois, Leena and Emmalyne. Tar-Mons went ahead of them, accompanied by his own security detail of three. Hames was surprised to see Emmalyne. As she got closer he could see she was tall and calm. There was no tension or fear in her face. Her blonde hair was pulled back tight and her hazel eyes were clear and focused.

Tar-Mons stopped twenty metres from the square, his eyes searched towards his son and then back his captors. Tar-mons skin looked more drained than normal and his voice loud as he called out to the villagers. "Greetings to you all. I am sorry that this is the way we should come together."

Throughout the throng of the villagers there was not one person standing front and centre like Tar-Mons. They stood together as one. From within this assembly came the voice of the gatekeeper. "We are unpleased to see you, Tar-Mons. We have left you alone but you have repaid us by sabotaging our lives! Your boy and this man have damaged our engine! It must be fixed or we will be forced to relocate to the citadel!"

Tar-Mons took a step closer; at the same time Prana, who had been standing over Hames and Bas, also moved ahead a pace. Hames could see Prana's hand rested on a weapon. A rectangular shape was held in the brown leather belt under Prana's tan t-shirt. Hames first assumption was that it was a typical energy gun, at that size likely to emit a small stun blast. He then got a clearer view of it – the yellow glow out of its rounded edges made it unmistakably a destabiliser.

The three guards supporting Tar-Mons all held long staffs with energy delivery points at their end. These would be no match for a destabiliser. Hames remembered stories of the outlawed guns from the Cernunnos and Lugus Civil War. The force of these devices could be final and horrific.

Tar-Mons could not contain the stress in his voice. "I promise you that we will give you all the energy you need. We have much to spare. We simply need time to prepare a transmission line to your village. We will fix what these visitors have undone. As a sign of good faith I would like my son returned."

"You will get nothing, Tar-Mons," the gatekeeper shouted. "Nothing until your man fixes the engine!"

"He is not my man and I am sure you have already asked him to do this."

"He won't do it! He says he can't, but he is lying."

"I have met this man; I do not think of him as a liar," responded Tar-Mons.

The COO called out to Hames, her tone cautious. "Can it be fixed, Chief?"

"No, I made it so. It's done. They were using an old style particle conversion drive. They're the reason our repair engine blew out. They're the..." an elbow from Prana to Hames' forehead finished the sentence of the chief mission officer.

"That was unnecessary," the COO called out as she paced back and forth behind Tar-Mons and his three guards.

"We don't want to be connected to you!" a voice from a female villager called out to Tar-Mons.

Another voice hollered to the lead administrator, this time a young man. "We don't want your harmony or your mind medicine. This is all a trick to draw us into you!"

The gatekeeper seemed prouder and stronger than ever. "Do you hear them, Tar-Mons?"

"I hear them!" The voice of the lead administrator seemed bereft of authority, there was only fear for his son.

Joseph Prana yelled out, "We don't want your transmission line! We want our engine fixed, or a new one built. Its power kept us untainted by the false feelings of Universal Direction!" Loud cheers rang out amongst the village population.

"I understand, but I can only do what I can do. Perhaps we..." Tar-Mons was interrupted by a flicker in the illumination of the main cube. "Perhaps we could..." Again, a flicker and another. On the fourth flicker the light in the main cube surged for a moment and then it was gone. The last second of vivid brightness served only to highlight the darkness that fell around them. Hames heard panic everywhere around him as the fearful and angry voices of the villagers called out for the light. Bas immediately got to his feet and pushed through Prana, knocking him to the ground. Hames, in a less nimble manner, also got up and ran for his crew. In the dark he immediately tripped over some of the chairs that surrounded the main square. Screams bounced of the wall of the cube, completely disorientating him. He got up and moved forward again, but was dazed and unsure where he was going. Bodies brushed and bumped him as he moved further away from his starting position, finishing on a side wall.

The lights within the main structure pulsed as they slowly warmed up again to a limited luminescence. The main cube was now lit like a single candle in a black room as all the other smaller cubes remained darkened. This was enough for Hames to make out the scene ahead of him. The COO and most of the *Medens* crew members were holding hands looking around for someone, hopefully him. Tar-Mons and two of his guards were linking arms, with Temett, the *Medens* second pilot, by their side. Bas-Mons was only a few metres away from his father. The cube pulsed again and lit up a further shade of brightness. The panic continued and the villagers, now better informed of their surroundings, ran back from their misplaced destinations.

In the crowd and chaos stood Jospeh Prana. He took aim at Tar-Mons and shot off his destabiliser. It crackled and broke in his hand, causing Prana to cry out. However, the damage from the device was done. Out shot a ball of energy that surged over the shoulder of a tripping Bas-Mons to hit Tar-Mons' first guard. The chain reaction of this tragedy was so quick it was hard to tell it was real. The first guard dropped to a puddle of liquid and other products that had formerly been his body. The energy from Prana's gun continued and surged back into the second guard and Tar-Mons. There was enough time for one last frightened look from father to son as Tar-Mons also dropped to the ground in an unrecognisable form. Hames saw in a split second that Temett would also suffer this fate, and it was done. The second pilot for the *Medens* was no more.

Hames was too deep in shock to respond. He knew that the pain he felt would also be experienced by the other members of the crew. He lifted himself off the wall and ran for the gate to exit the structure. His colleagues also ran to depart the disarray. As they crossed the final stretch of grass he joined their last few strides. Leena was at the front of their pack, looking back at Hames. Tears flicked from her face and her fists were clenched tight. The anger directed towards him was unmistakable. Emmalyne was running behind her shoulder. She wasn't crying but looked solemn as she rubbed Leena's back when they came to a halt at the exit. Hames turned around to see Francois and Siggy coming up, both panting. Francois leaned over, his hands on his knees. The portly Siggy Conson held his hips while walking in a circle. He was unable to process the startling events, and babbled, "They killed Temett. They killed Temett? Did it happen? Where did Temett go?"

Francois leaned upright. "He is gone. Gone from us yet still here."

Siggy was still unable to accept the reality. "Gone? Maybe he's back there in the dark somewhere." The ship's manager began a return run towards the chaos of darkness and sporadic fire of energy shots. "We have to go back for him!"

Leena spun from the main gate that was refusing to open, her wet face catching the light, revealing her distress. "He's gone!" The volume of her scream stopped Siggy in his tracks and made Hames flinch. "He's not here anymore. These crazy people killed him! All because of you!" Leena dove at Hames and smashed her two fists into Hames' chest, dropping him to the cool grass. "If you'd just stayed out of other people's business! If you'd just think of someone else apart from yourself this wouldn't have happened!"

Emmalyne held the small woman back, grabbing her dark brown arms and pulling them away to her sides. "Not now, Leena. We have to get out of here. They've still got that destabiliser gun back down there somewhere. If they get it working again, we'll all be gone."

Leena wiped the tears from her face and away from her neck where the water ran freely down the large golden circle spot on her neck. "I may not be as old or wise as you, Chief, but in my short life you are the most selfish person I have ever met." Leena withdrew herself from Hames and walked back to the gate. "So bad, so bad."

Hames himself was in shock. Temett was dead. It was horrible. He hadn't pointed the gun though. It was Prana. These villagers had caused the deaths of Xavi and Chung, couldn't Leena and the others understand that?

Having let go of the small first pilot, Emmalyne was free to resolve their situation. "Chief, pull yourself together and just stay clear of Leena for the time being. Francois, we need to get this exit point open." Emmalyne and Francois went to work on the gate and again she acted with the calm strength Hames had seen when she entered the cube village. "We need the gate to recognise our biosignatures, to let us exit. Francois and I will work on getting us out of here." Emmalyne looked around intently. "Where's the COO? She was right behind us."

Siggy slowly began to understand their situation and gazed back to the dark area of noise and movement. "Normally I'm the slowest one at this running thing. I can't see her."

Emmalyne was thinking of everything. "Siggy, where are those masks? We'll need them for the ride back to the citadel."

The ship's manager ran over to a far bench. "Still here, everyone take one for the outside." Siggy handed out the protective breathing masks the *Medens* crew had obviously worn down to the village of cubes. Without his carbon farmer suit Hames took a mask for himself; perhaps it had been

Temett's. His moment of introspection was disrupted when the COO came up with Bas on her back and a massive cut on her left cheek.

"That big guy with the gun won't be coming after us for a while, I clocked him good. Bas was knocked out by one of the villagers; we have to get back to the tower."

Francois turned from working on the gate. "What about the last of Tar-Mons' guards?"

Melini kneeled to catch her breath, Bas-Mons still on her shoulders. "The last guard stepped in front of us while I picked up Bas. Just as I ran I heard him scream." Her face dropped. "I don't think we should wait for him."

Emmalyne was working hard and in no distress or shock. She was a picture of pure determination. "Nearly done. And...done!" The entrance point morphed open and they all ran out putting on their masks. The COO rested Bas on the hover-cart while Emmalyne finished working on the gate to close it. "They won't be opening that anytime soon." She jumped onto the cart to sit with Bas-Mons. The young man was coming back around to consciousnesss.

Francois took the front driving seat and led them away from the village. The hover-cart pushed hard to reach the top of the rocky outcropping. Earlier that day Hames had looked out from this point to his first view of the illuminated cube village. Now, it no longer shone brightly; the cubes flicked in and out of focus. Bas held his head in his hands while Emmalyne rubbed his back in consolation.

The COO nudged Hames in the stomach and pointed back to the village below. "See down there? To the far side, one of the cubes is opening up."

Hames couldn't see at first as all the cubes sporadically shone and provided only the dimmest of views. On a second look he could see she was right. "There's a few hover-carts coming out. They look big."

"You're damn right they're big." The COO strained her neck to improve her view. "That one at the back has something large on it."

Francois turned to look then shouted back to them, "That is a weapon of some kind although it's hard to tell from this distance!"

The COO looked at Hames while the hover-cart moved over the crest, a look that was a mix of complete anger and utter disgust. He felt this from her without any words. Then Melini Guerra addressed the remainder of the hover-cart passengers. "Whatever that big device is, we've got to get back to the biotower to warn the others. Then we leave. For Earth or further onwards to Mermai, we're just leaving."

Hames couldn't agree more. As had happened thirty years ago, he was to depart New Galapagos in a hurry and under clouded circumstances.

Chapter 26

Fall of the Sun

THE EARLY EVENING SKY SET ACROSS New Galapagos as the remaining *Medens* crew raced the long journey back to the biotower. The wooden hover-cart was nearly drained of power, struggling to carry their weight over the concourse of cobblestones to the front gate of the tower. They arrived at the large white wall and the cart finally faltered, dropping slowly to the ground. Melini alighted quickly to warn the guards of the tower. Bas-Mons joined her to confirm the grim news. A moment later the large white wall distended to allow a space for entry and the *Medens* crew rushed through to the open courtyard and the thick tree-like pillars that supported the base of the biotower.

A low base alarm began to beat like a drum, causing citizens and administrators to scatter and look about in confusion. The *Medens* crew hastened towards the start of the winding pathway that led up to their quarters and a group of forty or so men and woman armed with protective gear and defensive artillery pushed past them. The defensive team gathered behind the front wall as they urged all the citizens to retreat to the tower and remain there until told otherwise.

The *Medens* crew knew they were in a dire situation. Melini, looking worn by the events, did her best to obscure her own worries. "Okay, we're going to head up to my room and let the defensive teams for the Mons Administration do their job."

Bas-Mons pulled away from the group and caught Quartermaster Chirila, who was calling out orders to the defenders. Hames assumed Bas was telling the quartermaster of Tar-Mons's death, which was proven correct as the quartermaster's face turned paler than its normal pallid white. The *Medens* crew left Bas and the quartermaster and continued their way upwards.

Emmalyne maintained her new calm nature and grabbed Hames' elbow. "Are you okay, Chief? We may need your to help get us out of here."

Hames like the others around him was in deep shock. He was caught off guard by her offer of reassurance. "I'm fine, really. I'm fine."

"Don't worry about Temett, Chief, he's with the Universe now, as always." Her voice was flat as if she was controlling any emotion she may have felt about Temett being shot and dispersed into nothing. She seemed so much stronger, more fit for this purpose. Whatever anxiety Emmalyne had been feeling before was gone. "Chief, can we get out of here, away from the planetoid?"

The base beat alarm continued its repetitive thumping, filling the humid night air and creating more confusion in Hames. "The *Medens* isn't ready yet for repair travel, only cruising speed. It will probably be safe though."

The crew reached the COO's room that had a balcony overlooking the bottom courtyard and front wall. Yolanda, Robero and Olisa joined them from their quarters.

The chief engineer was the first to seek an update, acknowledging Hames' presence. "I can see Hames is here safe and sound. Did Tar-Mons solve the dispute at that village?" No-one answered her question. Yolanda breathed out nervously, her rosy cheeks more flushed than normal. "What happened there?"

Leena was leaning with both her hands on a thin wooden table, her head looking down as if she was attempting to see a world other than the one she found herself in. At a barely audible level she let the remaining crew members know what happened to their colleague and second pilot. "Temett's dead, these people killed him in their village."

"Why? What happened out there?" Yolanda repeated her question, seeking answers from the room. Melini Guerra explained how a simmering dispute between the villagers and Tar-Mons had been ignited by a loss of the village's power source. She explained the deaths of Tar-Mons and Temett. Silence fell over the room.

Robero looked concerned and alert, standing with his arms closed and slowly nodding his head. "What happened to this group's power source?"

Hames' voice wavered. "They had a particle drive in the cube village. It was a rough, rudimentary drive that was adding to the mess of debris around New Galapagos. It's the reason the *Medens* failed. It's why Chung and Xavi died."

Looking to his friend Emmalyne for support, Hames found her expressionless face staring back at him. She was meant to be his friend, his colleague. Could she not see he was feeling bad about what he had triggered?

Leena's fists banged down hard on the wooden table, loud and in time with the base beats of the biotower alarm. Tears again ran down her face. "There was no reason to do what you did, Chief! It wasn't your decision to take! You should have waited. You only thought of yourself and not the crew!"

Hames' voice was incredulous, his strong arms reaching out to plead his case. "That's not true! I did think of you all. I thought of Chung and Xavi and all the people on New Galapagos that could have been caught in another Cataclysm."

Leena lurched up to Hames and stook directly in front of him, her small frame arched upwards, driving towards Hames' reluctant gaze. "You didn't know what was going to happen, nobody's that smart."

The small seed of guilt at the events of the last few hours was growing in Hames. However, the reach of this seed to sunlight and sustenance to be fully realised was restricted by his experience and knowledge of particle drives. He looked back at young Leena and could only half shrug. "It was luck, Leena. Just dumb luck that they hadn't blown themselves up. Not to mention the rest of the planetoid. I had no choice."

"There's always a choice!" Leena stormed over to the balcony to watch the events of below, wiping her face as she moved.

Yolanda was finding it difficult to support Hames. "Chief, what you did, it wasn't right. You went into someone's house and turned their power off. Without talking to them or warning them, or warning us. It's not good."

Leena was bursting with anger. "We've gone from repairing space to not even being able to look after ourselves. This isn't what we all signed up for. We must turn the *Medens* path back towards Earth."

Hames response was instant. "No."

Leena gestured vigorously at the crew. "There's too much danger here for everyone. For Xavi, Chung, and now Temett. We must turn around and return to Earth. The Organisation must hear about this."

Emmalyne offered calm, "Leena, I'm not sure Earth can do much about it now. It would be three years until anyone from Earth can get back out here. New Galapagos is on its own. So are we."

Suddenly, the walls around them shuddered like thunder coming up from the ground. Hames braced himself on a pillar, the broken compass from his gripping hand looked back at him.

"What was that?" exclaimed Francois.

Siggy Conson fell into Francois' back to stop himself from tumbling. "It felt like an earthquake!"

Francois, like all of them, was shocked. "It could be that device we saw on one of the hover-carts that was following us, Chief."

Hames was unsure of what he had seen. "From the look of the guns in the cube village it might be particle tech. They could be trying to destabilise the front wall, and the tower."

The COO steadied herself and focused their attention. "Everyone, look down on the main open concourse. There's the villagers' group and their hover-cart at the front gate. It's holding something large on it."

Hames squinted, unsure of his eyes at first. He was fast losing the trust of those around him so he wanted to be certain. A moment later the hover-cart shifted position and revealed its contents. "It's a particle cannon. It could bring the whole place down in a few more blasts or just keep shaking things up with small tremors. It's hard to tell from here."

Emmalyne stood tall and pointed beyond the cannon. "There's more carts coming up the concourse. I can see a load of people. I think it's more of the villagers."

Olisa, who had been silent until now, began to talk to Leena on the balcony, whispering messages of calm, connection and harmony. The young first pilot appeared to relax her shoulders and closed her eyes in some form of meditation. Francois joined them, reinforcing Olisa's message. "We are everything, we are still connected to Temett, he is still here."

Yolanda joined the efforts to examine the threat. "There's quite a few of them, about thirty plus huddling around that cannon and firing shots at the wall."

Hames could see the challenge for their attackers. "The villagers are too far from the tower to do any damage, they'll just shake it up. The wall, though, particularly where it shifts to let people in and out, well, that's another story."

The COO demanded more information. "Will they get through?"

"Yes, and soon, I'd say."

"Trouble," stated Francois.

"You bet," agreed Yolanda. "Chief, if they're using particle tech, isn't that going to damage the space around us?"

"No, it's not a particle conversion drive, it's more an attack on the very molecules of the object under focus. They're not converting space, they're trying to destabilise the wall's molecular structure, even for a second. Enough so it won't have enough strength to support itself and fall to pieces."

"I want to leave," Robero complained. "I am suddenly not feeling very well at all."

Melini Guerra held her hands to her head, lifting her hair up to reveal a face of immense confliction. "Damn it! We need to leave but we're caught up in this mess." She looked across all her crew. "The *Medens* coming here has caused this conflict. We'll stay until this is sorted, do our part." She turned to Robero, "You and Siggy head up to the shuttle platforms and prep the shuttle, just in case." Robero looked extremely grateful for this job. Siggy joined him in leaving for the uppermost point of the biotower.

Bas-Mons entered the room and stood next to Hames, both of them still marked by the dirt of the cube village ground they had been thrown onto by Joseph Prana. Bas was glassy-eyed and flushed as he glared at Hames. Seeing this and the potential for conflict, Melini Guerra moved over to Bas. "I'm so sorry for what has happened. Chief Hames was reckless in his action, it is very uncharacteristic of him." She turned her long brown-haired head towards Hames. He had not felt such a look of rueful disappointment since he was a child. "I can understand why he did this. It's not an excuse though."

Bas-Mons and the COO held each other's hands, a moment of physical peace as the biotower vibrated around them. "Thank you. I am not ready to even think about my father. We must defend the citadel and the tower above all else."

The COO let go of Bas-Mons' hands. "I'm afraid there is nothing much the *Medens* crew can offer you at this stage. We can wait and see it through."

Bas-Mons walked over to the balcony to survey the attack. "My father had guards and an armoury for a reason. Whether it was an attack by these villagers or something else. This is not the first time someone has tried to assault our Administration." He turned back to Hames, a wind ruffling his long curled hair. "They have always been seeking to attack our Administration. Until now, my father had managed the balance, until Chief Hames gave them extra motivation."

Hames felt Bas-Mons' simmering anger behind his quiet tones. The particle engine in the cube village was dangerous; he had made the right decision he told himself.

Another thundering shudder hit the biotower, shaking the walls around them. Yolanda was visibly counting in her head. "That was about six minutes I reckon. That's how long that particle cannon is taking to recharge between shots."

Emmalyne took a look at the cannon and pulled out a nearby access screen. "I'm reading that the tower's structural density is pretty good. That cannon may break wall down but it's too far away to inflict damage on the Tower, just like Hames said." She looked at him and he searched her face, hoping to find some sort of friendship or positivity from her. There was nothing. Emmalyne appeared more and more blank to him every time she spoke.

Yolanda moved to another side of the balcony, holding the curved white rail tightly. "The villagers have turned more hover-carts over, they're creating a barrier that will be difficult for the defensive fighters to breach given the repulsive power of the overturned craft."

Emmalyne spoke up, her hazel eyes fixed and certain. "A second particle cannon is joining the attack on the tower. It's just coming up now. One more shot from the cannons and they will destabilise the front entry point." She paused as the group all looked down on the chaos in the courtyard. "Right about, now." And, as if matching the verbal command from Emmalyne high in the tower, the front wall disintegrated into a pool of steaming hot plastic. This left ample room for the attackers to charge through and find targets for their handheld destabliser guns. A group in rough agricultural clothing pushed through the fresh opening. Without the restriction of the protective wall, the cube villagers commenced firing shots at the Administration defenders, the slightest touch of their destabilising weapons removing their targets from the world as they had done to Tar-Mons and Temett. The administrators did not appear to have a material defence for the particle conversion weapons. Some were now victims, reduced to a pile of nothingness.

Yolanda began to shed a tear, the first Hames had ever seen come from her. "It's horrible," she croaked. "Those people are dying terrible deaths."

Francois put both his hands on Yolanda's shoulders from behind her. "That's why particle weapons are prohibited. There is nothing remotely human about them."

Hames looked down at the cube villagers surging onto the concourse of the biotower, some nearly out of view. He saw Prana, the man who had struck him, leading a charge. He listened to the echoes of the ground level attack rise up to their quarters. They had come here as part of a mission to repair things, fix the fabric of space and here was the familiarity of a world being torn apart.

The front group of cube village attackers appeared to stall after passing the puddle of plastic mess that was formerly the front gate to the tower. They were at last being held back by some of the Administration defenders. Then, the COO noticed something the others hadn't. "There are two groups of attackers down there. The main ones pushing forward to the bottom section of the tower and a second group holding back behind the cannon, they look like a reserve force."

Yolanda agreed as she pointed downwards. "You're right, there they are. Staying back as if waiting for something. Can we zoom in on them from a tracking screen?"

Bas-Mons moved to the desk in the COO's room and an access point moulded itself out of the table. "I'll try to see if we can't get all the feeds for looking at this mess." In a few short moments Bas-Mons had brought up more viewing screens from the desk in front of him. They could all vividly see the carnage in the courtyard area where the cube villagers were cutting down the defence teams. The front line of guards for the biotower were dying from a dismorphing of their bodies into obscured versions of their former selves. Some shots from the defender's guns were hitting the odd attacker, but this was only slowing the onslaught.

Francois was sweating and spoke in his deep calm French-Cernunnosian tones. "It is not acceptable."

"Bas, are we in trouble here?" asked Melini.

"I don't know. We have more defenders coming, we should be able to hold the villagers off."

The COO looked down to the bottom of the tower then back at the tracking screen. "The second group have moved; they aren't heading up the central corridor, they're trying to access the paths to the left through the mountainous backing of the biotower."

"Why are they doing that, Melini?" Leena asked the COO, who could offer no answer.

The response did come, but from Bas-Mons, whose previously tight and alert face took on a look of horror. "They're targeting the sun-ray! The left pathway leads up to the sun-ray!" He looked shocked, as if his world was about to end. "That's the only thing of importance from that path."

"How many defenders are there for the sun-ray?" demanded the COO.

"A handful. Not enough," frowned Bas-Mons. "I must get there and protect it!" He ran immediately out of the quarters.

"Bas, wait!" Leena turned from Olisa and ran after Bas-Mons.

"Leena, stop!" shouted her cousin.

"Melini, we have to help them! The whole citadel depends on that sun-ray. We're responsible here!" Leena pushed the base of her palm across her Sentir spot, removing the sweat and tears from it.

The COO stood there with hands on her hips. "Cousin, the Chief did wrong but like Bas said, these people have clearly been waiting for this for a long time. It's not all Hames' fault."

"Whoever's fault it is we have to help them. We can't leave knowing this is happening."

Melini Guerra looked around at her crew. "What do you all think? Should we help Bas?" Clear agreement came from the determined faces of Yolanda, Francois and Leena. Hames felt disempowered to implore them otherwise. Yet, there was truth in the accusations levelled at him by Leena and his compliance in this task might go some way to mitigating his misdeeds.

Emmalyne considered this request as if it was a suggestion to walk along a quiet beach. "Yes, why not?"

"It's decided." The COO clapped her hands and grabbed a nearby satchel, swinging it over her back. "We'll head to the sun-ray and help the administrators save it. Whatever happens there, our time on New Galapagos is done, sun-ray or not."

The crew began a sprint to catch up with Bas-Mons. It took them eight minutes of weaving through long corridors and tunnels, some sloping downwards, some upwards, as they ran towards the sun-ray station based in the mountain supporting the biotower. Olisa followed far behind, her age limiting her efforts.

By the time they reached the rocky hollow that housed the station, fighting had already started. Eight villager men were already firing on the Administration defenders and the large blast radius from the destabliser guns was crippling the stricken protectors of the sun-ray. Bas-Mons was amongst them, battling alongside his compatriots. The Administration's pulse weapons were slow and indirect. Fortuitously, due to their ill-informed navigating of the paths of the biotower, the *Medens* crew had come in to an entry point that placed them behind the attackers. The cube villagers were atop and around the sun-ray equipment, some taking action to damage it while others fought the defenders. With their backs to the crew of the *Medens* and distractions around them, the village attackers were unaware that they had been flanked.

In silence, Francois' large hands grabbed two of the attackers at the back of their group and slung their faces into the ground. This unnoticed act was followed by the squashing of another cube villager's head with a two-handed slam into the rock wall of the hollow. The COO, Yolanda and Emmalyne all grabbed and held an attacker with the final two taken out by Bas-Mons and the Administration defenders.

Bas approached the crew with a large smile while Francois subdued the other attackers. "Thank you, my friends. I will have to learn your pathways in the tower, you seem to know them better than myself."

Next to them were the large and complicated mechanics of the sun-ray – different sizes of metallic boxes and panels stacked upon each other. It sat over two metres high and wide; a seat at the top was positioned in front of the main control screen. At its end, long cords of power and information stretched out to a large four-metre radius dish. The circular device held a rod in its middle that pulsed its beam up to the circle shape in the sky, the shape that was created by its partner station over in the Impenetrable Valleys.

"This is so essential to life on New Galapagos," said a relieved Bas-Mons. "We have some redundancy over in its partner station. If both were to fail for any great length of time we would suffer immeasurably." Bas-mons patted the side of the sun-ray device. He paused and began to feel around it's edges. "There's something wrong. It's overheating." Bas started to search each section of the structure.

"Yolanda, light drive tech and sun-rays?" The COO asked her chief engineer for assistance, hoping that there was some similarity between light alignment drives and the creation of an artificial sun.

"I'll have a look, no promises." Yolanda bounded up the boxes to the control seat, with Emmalyne joining her. She started to look through the cause of distress.

"They've done it. One of the battery packs is overloading." She faced Bas-Mons. "I'm sorry but one of your packs is going to overheat the sun-ray. You'd better get any replacements together soon." She shook her head. "I don't think this battery set is going to last. I reckon pretty soon it's going to combust and take other parts with it."

"The spare packs are on the other side of the planetoid." Desperation was plain on Bas-Mons' face. "We should be able to power this down. Can you fix it?"

"I'll do my best." Yolanda started work on programming the station. "I'll buy you about half an hour, maybe."

Bas-Mons jumped off the sun-ray machine and rushed over to the COO. "I need to get up to the shuttle platform, get over to the other station."

"Let's just give Yolanda and Emmalyne a minute to make the fix then we'll leave."

On the screen in front of Yolanda the battery pack lit up red and then turned to brilliant white, as did the whole control screen, illuminating the sun-ray cavern.

The chief engineer covered her eyes. "I got it wrong!" Yolanda then pushed Emmalyne off the top of the machine, the statuesque young mission officer tumbling to the dirt. From underneath Yolanda, half the boxes of the sun-ray machine combusted, destroying the wall behind it and throwing dust and rocks into the air.

The small blast was enough to throw Hames flat on his back. His head rang while stones and pebbles flew around him. He pulled himself upwards and shook the debris from his face. Through the dust cloud he could no longer see Yolanda. Amazingly the cords from the sun-ray remained in place, still connected to the large dish pointing out to the sky. Operationally sound yet diminished, the sun-ray continued to shoot its beam outwards. However, the dish was now altered in its angle. No longer did it fill the circle in the skies above New Galapagos. It shot beyond the planetoid, out into the dark of space, finishing its stream at the fourth moon. Even through the haze of dust and confusion, looking out of the stone cavern to the stars, Hames could see the fourth rocky satellite of the New Galapagos planetoid begin to emit a glowing green light.

Chapter 27

Dover Docks

NISHA LAY STILL ON THE CORRIDOR floor, her dark blue jacket lying open in the thin layer of liquid. The cold water had seeped into her grey pants and boots, adding texture and weight to her apparel. Awaking slowly she looked around. Nisha groaned, realising what had happened. "Oh, the corridor theme. I slipped. Perfect timing, Nisha." She pulled her damp body off the floor. Her shoulder throbbed with as much pain as it had on the day after the bomb blast at St Paul's. Curse these themes. *It will be a fun way to learn about Earth*, she thought sarcastically. *Thank you, Hames.* She winced and breathed heavily as she walked forward. Jasiah was nowhere to be seen.

She tried the communication link on her black bracelet. Nothing happened except for the time flashing back at her; the device was in reset mode. There was no record of Nisha Pinto. It was as if her profile had been removed entirely. Her heart raced and then skipped to the panicked thoughts flooding her mind. "Damn it!" How long had she been unconscious? Her communicator said the time was 7.10 pm; not long until the *Python*'s 8 pm flight to Mermai. Could it be true what Jasiah had said about her profile being stolen? Could the *Python* be attacked tonight?

She thought of her boss, Peder Jonslow. He could search for someone using her profile and biosignature. Peder should be able to find them by looking for her employee number; it wouldn't take long to stop them. Instinctively, she began to move for the exit and out towards the departure for Dover. A sky transport arrived immediately, presenting her with the first amount of luck she had received for some time.

The small ship rushed out of London, reaching green fields below while the skies darkened with the approaching night. As the transport moved towards Dover she realised how bereft she was of resources, solely equipped with a broken communicator, wet clothing and a furrowed, frustrated frown. The few passengers around her shifted away and looked disdainfully at her vagabond appearance. Her resentment at this situation was immeasurable. How was it her role to stop the *Python* from leaving? Perhaps this wasn't even required. Jasiah could have been wrong about what he heard. Maybe nothing was going to happen to the *Python*, at least today. She tried to access a communication point on the transport craft with no success.

Exasperated, she opened a small window on the mid-section of the transporter. Air surged through to her damp brown hair and drops of water flew from her green-skinned face to the floor. Refreshed, she moved to speak to her fellow travellers, to ask for help. Again, the small number of passengers recoiled. At this point in time she was not an attractive individual to be around. What aggravated her most was that someone had stolen her information. Someone from Montaxe-Deus had her profile and could be going to the shuttle docks tonight. How they planned to fake a biosignature was beyond her. This group must be confident and bold. She had heard that on some worlds identity theft was punished by cryo. It was such a foolish idea, she almost felt sorry for them.

Alone and cold Nisha felt her stomach turn in knots. She wished Hames was with her. He could fix her communicator, he could hold her and keep her warm. Instead she was alone and waiting to reach Dover to warn them some strange group was trying to sabotage the *Python*.

The wind lashed her face again. "This is not my problem," she whispered to herself. The warmth of a tear filled her eye, its temperature quickly equalising with the rest of her ice-like outer shell. Soon she would offload this issue when she arrived at the shuttle docks. She would be there in minutes, leaving enough time to stop the flight and make this whole situation safe.

As it had done many times on her travels to work, the small transport craft jerked to the right and flew downwards, positioning itself to fly across the dark night water of the Channel. It moved low towards the arrival point of the shuttle docks complex then flew into the sea-level opening and unfolded to allow its passengers to disembark.

All those onboard departed in a flurry, taking lifts up to the departure lounge. Nisha was left alone with only the waves that lapped the hangar opening to keep her company. She ran to a far wall and accessed the nearest communication terminal. Immediately, Peder Jonslow answered the call. His large forehead lifting in surprise.

"Nisha, what are you doing here so late?"

Filled with tension and adrenaline, Nisha had to calm herself before she could speak. "Peder, something terrible might be happening tonight. I think someone is planning on sabotaging the *Python*. Something to do with its 8 pm flight to Mermai!"

"What? Don't be ridiculous!" Peder's blond eyebrows pushed downwards in deep disbelief. "The security systems would have picked this up ages ago." Her boss looked away to another screen and spoke again. "The *Python* returned at its 6 pm slot and there were no problems whatsoever."

"No! You're not listening to me." She breathed quickly again, matching the urgency her body felt. "Weird things are happening lately, there's all these groups positioning themselves, meetings and pamphlets," she tried to breathe and talk at the same time, but it was hard. "There's been bombs, fighting at a healing centre and..."

"Nisha, you're not making any sense."

"Jasiah, my cousin, he was with a group that the Agency were following. Jasiah believes someone is going to target the *Python* and damage the ship as some sort of attack on Universal Direction!"

"Okay, okay. Just calm down. What do you think is going to happen?"

"I don't know. They have my details, they scanned me. I think they're using me to get access to the *Python*."

A large frown again appeared on the face in front of her. "Nisha, how did this happen?"

"It's a long story. I attended a meeting of this group that Jasiah is a part of."

"Who is this group anyway?"

"Montaxe-Deus, the Agency knows all about them."

"I have never heard of them, they are news to me."

The calmness and lack of action from her boss was beginning to frustrate Nisha greatly. She fell back onto process and protocols that she knew in detail. "We have to evacuate; it's an emergency, Peder. It's a trigger for the emergency management plan. Activate the EMP now and get people out of and away from the *Python*!"

Peder Jonslow pensively looked down and then back out through the communicator screen to Nisha.

"Right, I'll sort it. We'll evacuate the space docks and the *Python*."

"Thank you. Thank you for doing this." Relief washed over Nisha and she felt warmer than she had a few moments ago.

Peder shook his head back at Nisha. "People won't like this, Nisha; it's a lot of distraction."

"It's the right thing to do."

Her boss paused some more. "You are correct as usual. It's the right thing to do."

"Great. Let me in. I'll fix my profile and give you the information I found."

"No, no, no. You have done enough. We can take it from here. We'll start evacuating people now. Go home and start packing. You leave in two days, yes?" Peder Jonslow was giving her an order and she was happy to take this direction.

"Yes, and thank you again."

"Safe trip, Nisha."

Nisha placed her hand on the terminate button. "Good luck with it." With a lightness in her body she started walking back to the unfolded transit craft that awaited the next round of passengers returning to London. She was done and there was no way to articulate how good it felt. Her shoulders relaxed and she felt a sense of achievement. Closing her eyes, she concentrated on breathing, in and out, in rhythm with the flow of water outside. It would be a moment of peace until the lift doors opened, expelling a voluminous crowd of evacuated passengers all clamouring for a safe exit.

Any second now the emergency plan will kick in, she thought to herself. After another minute nothing happened. Slowly the feeling of tension returned to her chest and she looked back to the access panel to check the status of the shuttles arriving and departing. None were grounded. What was displayed, however, was a warning message across the bottom of the departures screen, a warning about her! It read to avoid Nisha Pinto. To take caution and notify security. Nisha considered the words; they were surely designed to help capture the person who had co-opted her profile. Then, in shock, she saw her photo posted alongside the warning message. She was the enemy after all.

"There must be a mistake," she said out aloud. "Peder is meant to evacuate the *Python*, not try to capture me!" Nisha tried to raise her boss once again. No answer was forthcoming. She immediately felt responsibility for the situation flooding back onto her shoulders. Without her active profile, though, she would not be able to reach her offices. There was only one choice – she must get up to the shuttle departure lounge.

The lift opened and carried her straight to the top of the Dover Docks with its gates and shuttles destined for Earth's orbit. Despite the encroaching evening it was both busy and brightly lit. Even in her panic she could still appreciate the beauty in the curved walls and views across the sea. The activity around her appeared disturbingly familiar and normal. Passengers were walking in and out of departure gates, some returning from space, some leaving for a trip far away. There was no evacuation.

Ahead of her she could see people lining up at Gate Four and the shuttle for the *Python*. As to be expected, many looked to be Mermain and bound for their home planet. The security scanner was manned by one of her recent inductees, Olive Sassonet, who was processing passengers for the shuttle.

"Olive!"

"Oh, mam, I did not see you there!" Olive looked surprised at Nisha's bedraggled and wet appearance.

Nisha pushed herself in front of the passengers lining up for the security scanner.

"Olive, have you been told of the emergency?"

Olive Sassonet showed no comprehension of what was going on. "No, mam. There is no emergency as far as I'm aware."

Nisha pointed firmly in front of her body so as not to alert the passengers behind her. "There is an emergency. Let me tell you there is!"

"No, mam. I'll...I'll double check my communicator." The young shuttle docks officer flicked through the screen of information on her wrist. "Mam, it says nothing about an emergency. The only warning is...it's about you." The officer looked back at Nisha, seeking reassurance. "It says if we see you or catch use of your ID, the person must be detained. It's from Director Jonslow himself."

Her stomach felt hollow. Peder had not gone ahead with the evacuation. Was he trying to stop her or stop the person who was using her profile? In that moment Nisha was resigned to her duty, an obligation to prevent any danger to the people on the *Python*. She reached into her pocket and fumbled through to grab a navy hair band. She pulled her hair back tight and secured it with the band.

"Olive, do you remember how we tested you in that last induction session?"

"Yes, we answered many questions about the particle conversion drive hub, which ships left from what gate, that sort of thing."

"And what about making a choice to safeguard people? We spoke about choosing to protect people." Nisha motioned her head to all the people around them, including the passengers behind her, starting to grumble about the delay.

"Yes, you suggested we make a choice to protect people's lives rather than stick to our set roles."

"Olive. I am asking you to make a choice. I am trying to safeguard a lot of lives. You can see that in my eyes. Help me, Olive. Help me save these people."

"I'm not sure. The warning message says not to engage with you."

"I can't explain that right now. Look, I'm telling you, Olive, I know something is terribly wrong and I have to act. I can't stand by and let someone else fix it, or people's lives could be lost. Let me go up to the *Python*."

Olive Sassonet looked back at the line of people growing and then at the sincere and strong stance of Nisha Pinto.

"Oh, okay, yes."

"You've done the right thing, Olive."

"It's Olisa. And I hope so." An unsure-looking Olisa turned to usher in the remaining passengers to follow Nisha into the waiting lounge.

A short time later they had all boarded and the shuttle took off towards the space docks. While it surged through the night-time clouds there was an interlude for her to process what had happened. She banged on an arm rest. "Peder knew I was leaving in two days' time." Peder was part of this trouble. After not activating the emergency plan he had also deemed her a public enemy. Nisha didn't know why this had occurred. All she knew was what had to happen next. She would have to evacuate the *Python* herself.

The shuttle arrived at at 7.52 pm, only eight minutes before departure. The ship would soon be full with over one hundred and sixty passengers. Nisha exited the craft and began running up the curved escalator to the gate. Her diminutive body weaved in and out of fellow passengers; single people and couples holding hands or families walking side by side. Some hauling luggage and some with simple backpacks. She ran past them all. Some were happy and content. Others just stared blankly into their immediate future, temporarily in transit. There must have been hundreds of people coming and going, challenging her as she ran to the entrance to the *Python*. Challenging her to save them all.

Nisha reached the gate and there were no security guards to be seen. *Peder must have removed all the officers on the space dock*, she thought to herself. It was fortuitous, though, as this made it easy to run up to the gate and onto the *Python*. The ship's welcome officer, a middle-aged Mermain woman with long auburn hair, didn't recognise Nisha as she skipped past the official boarding check.

Nisha knew her way around the *Python* and strode directly to the bridge. What to do? "Think, think, think," she spoke aloud. As she approached the door it opened for one of the *Python* pilots to exit.

A petite woman in uniform unfastened a side compartment and saw Nisha. The officer showed a toothy grin and then frowned. "You're that docks officer under Peder Jonslow! I'm..." The girl turned back to close the door to the bridge but not before Nisha pushed in with her. "Hey! You can't come in here! Regulation 17 states that only pilots and navigators are allowed on the bridge!"

A dark-skinned man in a pilot's uniform jumped up quickly, but was dragged back down as a protection belt caught on his large stomach. "I'm calling security!" Nervously he fumbled around the console, sweat immediately beginning to pour out from his grey thinning clump of hair.

"Don't bother, they won't be coming! There is no security," said Nisha.

The sweating pilot finally unbuckled himself. "I'm the captain here and I say what happens. Get off this ship!"

Nisha relaxed herself and spoke tersely. "Gladly! You also have to get off this ship! Someone could be sabotaging the *Python* tonight – they probably already have!"

The young woman in uniform looked back at Nisha, confused. "Isn't it, um, you? Aren't you the security risk?"

Nisha noticed that on either side of the young woman's neck were two dark navy spots. She instantly understood what role she performed. "You're the navigator, right?"

"Yes. Appointed and empowered under statute 16H8," said the navigator, blinking back at her.

The raised blue spots were small yet hard to miss. These linkage points would give the navigator control over all the *Python*'s systems. "Well, then, you need to connect with the ship and call out an evacuation. You need to power down the *Python*."

"But why would we do that?" responded the navigator.

"You have to leave and evacuate the ship." Nisha knew enough about how particle conversion ships worked and not just as a shuttle docks officer. Having an interstellar travel expert for a partner helped her pick up a thing or two. "You can't start the particle engine spinning up, it could be compromised!"

The young navigator blinked at Nisha. "It's too late. I did it just before I saw you. We'll be ready to leave on time."

"Stop talking to her!" the pilot demanded.

"But what if she's right? Check the security feed!"

The pilot sat back down in his chair. After trying various screens he sounded confused. "I can see the feed to the outside. There's no security guard at the gate." The pilot tried to activate a communication panel near the video feed to no avail. "Its not working, is it?"

"No, it's not!" Nisha shot back at him.

The navigator pulled two blue cords from a panel to connect with her neck. She stared into the distance as if reading a book. "He's right, you know. We've been cut off from the outside. We can see out, communicate out, but nothing is coming in. I can't connect with anything else but the *Python*."

On a screen near the flight controls a low tone sounded to draw their attention. The greying pilot pointed uneasily at the panel ahead of him.

"The docking mechanism has been shut off. The gate manager has sealed us in. That's a bit early."

Begging now seemed to be Nisha's best option. "Please! Finally get this. You can't go. Someone is trying to destroy the *Python*."

"I'm contacting the gate manager to sort this out." The pilot pressed a communication link with no success. "The comms link, it's still down. We really can't get out to anyone."

The young navigator was nervous now; she started to quote more regulations at the captain. "RG161 states that in the loss of communications the ship is to move to secondary protocols of deep space messaging."

"Yes, yes," said the pilot. "Keep sending out the distress feed. Hopefully they'll catch it somewhere else and this can all be resolved."

Nisha felt like she was getting somewhere. She stood with her hands on her hips. "Do you get it now? You can't move off."

The pilot begrudgingly agreed. "Fine, fine, let's disembark every passenger, once we have the gate unlocked. Are the internal communications working?" The navigator looked away then nodded back a confirmation to the pilot. "What was the welcome officer's name?"

"Runa."

The pilot seemed to be calmer and less sweaty. "Runa, no need to worry, but we have to evacuate the *Python*. A routine check has found an error preventing take-off. Can you please start to disembark the passengers?"

"Yes sir. Will we be boarding again?"

"Thank you, Runa, that is all."

Nisha viewed the screen in front of the bridge. It was only minutes before 8 pm. The ship was getting ready to trigger the particle conversion to move them through the black of space in the blink of an eye. Somewhere, something on the ship was going to try to stop this from happening. She had at least made it to the bridge to alert people of this danger, despite the obstacles that had come close to halting her progress. It appeared the only obstacle left was getting the pilot and navigator to move quick enough to evacuate the ship.

The communicator came back online from Runa. "I can't open the gate. It's sealed on the other side of the docking mechanism."

The pilot looked around then pointed a finger at Nisha. "You did this!"

"Why would I lock myself in here? Wouldn't I just do it from the outside!"

"I d-don't know!" the pilot stammered.

"Look at my face damn it! Do you think I want to die?"

"No, no you're right."

The navigator quivered out more regulations. "RG53 says in the event of suspected or actual sabotage the ship must return to its last known port of call."

"You mean here. Earth? We're here already!" The greying pilot shouted back in frustration.

Nisha took control of the situation. "Okay, you have to stop the engines now! Shut down the ship!"

The pilot sat back down in his seat and for some reason buckled up.

"You're not taking off!"

"I know, I know, it's just a habit!" The pilot was drenched in his perspiration as he pushed controls and entered codes that didn't appear to be accepted. "I can't shut it down! The engines aren't responding!"

Anger and energy pumped through Nisha. "Get down to the engine bays, now!"

The pilot fumbled to unbuckle his restraint, his hands shaking, and left for the engine bays.

The navigator commenced working through checks in her head as she remained connected with the *Python*. "The ship is locked and working up for a departure, at 8 pm sharp. That's only two minutes away." Panic then reached her eyes with her latest realisation. "The logs have been wiped! We're being sent to nowhere!" The young woman was distraught.

The internal communicator came online and the pilot spoke. "They're dead! All the crew in the engine bay are dead! I can't turn the engines off! We're going to convert space and there's nothing we can do to stop it!"

"There must be something you can do? Stop the ship from here? We can't just randomly convert space to an unknown point!" Nisha pleaded.

"I'm sorry, I've run out of options. There's no route and the engines are locked out!" the Navigator began to sob.

It didn't feel real. Perhaps if it did Nisha too would be in a flood of tears. Only a few hours ago she was happily packing for Mermai. Getting ready for home. Now she was stuck on a ship about to venture directionless to the depths of the Universe. Would they stop or would they hit a star? Would they even know where they ended up? She looked out the window down at Earth. Its blue and green curves shone against the void of space. It didn't look so bad after all her recent complaints.

She tried her wrist communicator once more. Turning away from the navigator she whispered his name. "Hames. Hames, are you there? If you get this, ever. You must know I love you, always. Don't regret anything, the past is past. Be safe and strong my love. You are everything to me." She allowed herself one tear and one exhale of sorrow. Nisha reached to the pendant of the mermaid Sovann Macha. She wondered if like the demon daughter of Ravana she too would end up lost and alone.

Nisha rested against a wall of panels and screens. It was over. Out ahead through the viewing windows was endless space. That was where they were going. Putting her hands to her communicator bracelet once more to try Hames she remembered the small metal square attached underneath. It was the blue data chip given to her by Bala Garang, Hames' instructor from the Agency Academy.

"Hey!" she shouted to the navigator. "This holds the travel coordinate routes to outer systems."

The navigator's eyes lit up. "Where did you get that?"

"I just have it. Use it!" New life breathed into Nisha although time was evaporating. The *Python* was humming with a vibration that suggested it was soon to depart. "Look at the screen, you've forty seconds to load the directions to Mermai!"

"That's not enough time!" The navigator inserted the data chip into a panel and frantically typed on the screen in front of her. "We can't do it!"

A counter showed twenty seconds to departure. "Come on! Even if you only get the first round of codes loaded it will be better than nothing! At least get us headed in the right direction!" Nisha pleaded.

"I'm trying, I'm trying!" More frantic finger taps ensued from the navigator.

The *Python* was now vibrating more than Nisha had ever felt before; this was irregular for particle conversion ships. "Hurry, hurry!" It may not be a lost situation after all; she still might see Hames again.

The screen counted down to four seconds and the navigator exclaimed, "I think I've done it! I think I got a route planned from your chip."

The screen ticked over from two to one. Despite the *Python* rumbling violently it did not move forward. Rather, a sense of energy filled the bridge and shook the ship, making it impossible to remain upright and steady. Nisha had to shout to be heard. "What's happening? Is this normal?"

"No, something's wrong. We're too late, I'm sorry!" screamed back the navigator.

Without warning a black and grey form of electricity crackled around the *Python*. "Buckle up!" Nisha jumped into the pilot's seat and clicked in, an action mirrored by the navigator. The grey lines fizzed around the outside of the ship, shaking the *Python* into a frenzy. Then it happened. The *Python* commenced its journey of particle conversion. Instead of a smooth and seamless departure it was with the sound of thunder and devastation. To Nisha, it felt as if the Universe had been torn in two like a giant stick broken across her knee. The noise escalated in her head as the grey electricity burst into black, taking her and the *Python* into the night.

Chapter 28

Running for the *Medens*

THE SUN-RAY AMPLIFIER BEAM CONTINUED to pour its skewed ray of light from the New Galapagos planetoid. The beam from its slanted dish no longer shone out to a circle shape in the sky; instead the rod of light stopped at the fourth moon thousands of kilometres away, igniting Verona in a vibrant visage of emerald light.

No longer did the false sun send down its warmth and luminescence onto the people of New Galapagos; an unexpected darkness fell upon them. Lamplights across the streets and buildings now served as the main source of guidance under a curtain of stars. The fourth moon with its new aura of shimmering green accented the streets and buildings with a greenish tinge.

Hames sat up on his knees and looked through the dust and rubble to where Emmalyne and Yolanda had been working on the sun-ray station. The strewn boxes and broken screens were evidence of their failed attempt to save it from sabotage. Despite the overpowering of one battery pack, the device was still continuing to function, although in a diminished capacity. Hames' first effort to stand up failed and he fell down to his bottom, his legs and arms dropping in exhaustion.

Emmalyne, who had been thrown away from the blast, showed resilience with her new-found strength. The junior mission officer pulled herself up to her knees and then fully upright. Emmalyne's recovery paused when she saw Yolanda's hands and head amongst the debris. The lifeless eyes of the chief engineer reflected the glow from the sun-ray beam. In her last moment Yolanda had recognised the imminent overpowering of the battery pack and pushed Emmalyne to safety. The contained explosion had decimated the station. In truth, nothing should be operating. It was a last testament to Yolanda's skill that the sun-ray was still working.

Olisa finally arrived at the stone hollow that held the devastation. Her elderly frame stood over Hames, peering inquisitively into his stunned eyes. Without a word she leant down and helped him up. Melini was now walking about. "Olisa, your med-tablet. Scan for heat signatures behind those rocks. Check if any of the New Galapagoans are still alive!" She stopped to her right to help Bas-Mons, who was tending to an unconscious Leena; blood was pouring from the top right of Leena's head after it had crashed against a rock wall.

Olisa struggled to breathe in the dusty environment. Her old hands pulled a soft tablet screen out, which she held in front of the mess of rock and metal. She coughed and turned blankly to the COO. "There is nothing. Yolanda is passed from this moment." She turned to Leena on the ground. "Leena has a slight concussion and should be able to move soon, no significant damage."

"Thanks." Melini appeared to gain a wave of energy, summoning strength to continue her responsibilities as COO to take care of her crew. "Francois, pick Leena off the ground, carry her if you have to. We've got to go. Time to look after ourselves." She looked around at Emmalyne, Hames and Olisa. "Any arguments?" None were forthcoming. "Right, let's get up to that platform and see if Siggy and Robero have got the shuttle ready."

Looking over the bodies of his fallen defenders, Bas-Mons took a solitary moment of reflection and began to dust himself off, his long curling hair flecked with debris, his body bearing cuts and emerging bruises. "I need to get to an Administration shuttle and help my people."

The COO spoke to him as if he were one of her own. "Bas, it's devastation down at the bottom of the tower; you may not be able to do much by yourself."

"I'll take a shuttle over to the valleys and the other sun-ray station. We have equipment and people who can help."

"It's your choice. If you could guide us to the top of the tower it would be much appreciated." The COO dusted herself off as Francois picked up an unconscious Leena.

Bas took a short step to the edge of the sun-ray cavern and looked down to the bottom of the biotower at the remaining attackers. He shook his head as the mountain cavity hummed with the vibrations from the attack. "There isn't much time. That particle cannon is having an effect now. I can't believe this is happening. We had no idea. They must have been waiting and preparing for this."

With his large thick hands Francois cleaned Bas' back. "Young Bas-Mons, you are doing amazingly well. Time will resolve all this; what will be will be." Olisa nodded encouragingly at this most appropriate use of Universal Direction conventions. Hames felt the discomfort from his bruises and scratches being compounded.

The platitude seemed to ease Bas' mind, introducing peace to his disposition. He faced the crew of the *Medens*. "Let's go. I'll lead you up to your shuttle." He ran out of the mountain hollow towards the biotower, and the COO and what remained of her crew followed with Francois carrying Leena on his back. They entered the white-walled corridors of the biotower that were rumbling as the particle cannon continued its assault. Cracks in the wide cavernous walls began to appear more frequently as they ran through the passageway to the main central corridor.

"Keep moving!" urged Bas-Mons, stopping to run sideways, allowing them to match his pace.

A thunderous crack shook them all. It felt to Hames as if the tower had split in two. But it was still holding together. The noise woke Leena from her slumber. She slid off Francois' back, assessing the situation, counting with her eyes the remaining crew members.

Emmalyne was aware of her friend's dazed and unspoken question and offered a shoulder for her to lean on as they ran further upwards. "It's Yolanda. She didn't survive the blast."

Leena shot an angry glance at Hames. "No Yolanda. No Temett. Time to turn back to Earth wouldn't you say?"

The tower continued to shake with dust as rocks fell all around them, foretelling its imminent ruin. Bas called back, "The tower is collapsing; it could fall at any moment."

The COO clapped her hands together four times back at her crew. "C'mon, c'mon, let's get moving!" They all increased the pace. Now that Leena was conscious and being assisted by Emmalyne, Francois began to carry Olisa. Hames would not openly admit it but the Emmalyne who had awoken weeks before from stasis would have been unhelpful in a situation like this. Unlike Marak Chakla, Curate-mente had seemed to enliven his young friend; it had given her a new-found resilience. Marak had lain there ever since the mind healing thirty years ago, a silence only broken by the images he had presented to Hames' mind. They were not far from the quarters of the bedridden man. Without thinking, Hames deviated from the path and ran down a side corridor.

"Hames, where are you going?" shouted the COO as she spotted his new direction towards the hospital wing.

"I'll be right behind you. One minute or so. Just get going and hold the shuttle for me."

Emmalyne called out, "Hames get back here!"

He flinched at this strong direction; this was Emmalyne in control of herself and assertive in her environment. His mind was set, though, and he separated from his crew once more.

As he wound his way to the hospital wing, he saw through a far window the sun-ray amplifier beam shooting out to the sky, still focused on the fourth moon. Despite the catastrophe all around him, the loss of crew members, he was most concerned with this. The fourth moon had been part of his life when he left New Galapagos years ago. It was the place Marak had been touched by the strange form of energy that was similar to that of the Ugor Eels, and their egiya energy.

Hames arrived at Marak's circular holdings. It was calm with only the slightest sound of disturbance. It seemed removed from the vibrations and violence of below. Unlike his last visit, there was a readiness within Hames to understand what was happening inside his old colleague. For the first time since the Cataclysm he recognised that an experience beyond the known world was required. Hames had always railed against the subjective and the supernatural, yet here he was, looking at the man who had touched him with something beyond his comprehension. A message that had sent him to the cube village and shown him the answer to the accidents on the *Medens*. It was unfortunate that the same message had set off a chain of events that led back to the room in the tower, which was being shaken towards its demise.

Hames pulled his med-bar from a side-pant pocket. In a few short movements he reset this device to scan rather than heal. He knew what he was looking for. The *Medens* had alerted them to the egiya energy when the Ugor Eels had appeared. When he reached Marak's bedside, the confirmation was instant. Eigya energy was still within him. Whatever had touched him all those years ago on the fourth moon was still there.

Without thinking, Hames placed his hand on Marak's chest to say goodbye. Then, just as it had happened days ago, like the feeling of awareness that had come upon him during the accident in the light alignment drive room, he saw an image in his mind. This was an image not of his own choosing. Hames saw a darkness that was only penetrated by a faint silver ball. A ball that flexed and then shrunk to nothing, leaving only the black. He removed his hand and was left with the memory of the instruction just provided. Small trickles of dust and plastic from the ceiling above floated down onto Marak Chakla's body.

It was time to leave. Marak, though, was plugged in to multiple feeding tubes and devices. Hames had no idea where to begin. It should feel wrong to leave a human here, one who had affected his life so much, but he had no choice. It would be impossible to move Marak and his support system. The shuttle would be at the top of the tower and he was already minutes behind the others. The decision to leave was practical and sound. "Goodbye. I'm sorry this was your life. But I guess I don't really know what your time has been like."

Hames ran from the hospital wing as fast as his body would take him. Even though the shrinking ball made no sense he was certain of its veracity. It was a matter of faith that the next answer would in time reveal itself.

After winding up the central corridor he reached the platform. Coming out at the top of the path and through a clear opening he was greeted with sparks, cables and chaos. A crowd of people were trying to board the two shuttles left for the biotower administrators. The large pentagon-shaped platform was overflowing dangerously with people pushing and shoving their way to the front of the line. Bas-Mons was at the back of this group trying to regain control. The shuttles' doors remained shut. Hames pushed through the back of the crowd and past Bas. Men and women in tan tunics with different trims and patterns all clamoured for the last remaining life boats. Elbows and bodies pushed backwards, knocking Hames about as he struggled past the edge of the group.

The tower was swaying and Hames felt like he was standing on a plate being balanced on a stick. Through the chaos at the far point of the platform he saw the *Medens* crew members watching the disorder from a vacant bay. He jogged over with his hands stretched out and asked the obvious question. "Where's Siggy and Robero with the *Medens'* shuttle?"

"They're circling above us. They went up there to stay away from the madness." The COO gestured to the group of workers from the Administration who were staggering in and out of time with the swaying tower, all of them arguing for entry to the last two shuttles. "We'll be lucky to all fit in our shuttle; it will be damn tight with eight of us. Siggy's reshaping it right now to accommodate us." Hames noted how easy they all seemed to be about leaving the New Galapagoans to their fate. While he was more open to the unknown wonders of the Universe, he would never be okay watching people pass from this world.

Leena was now standing upright on her own. "I'll call them down; time for me to fly us out of here." Pressing her communicator in the gold Universal Direction symbol on her brown wristband, the first pilot directed the shuttle to lower itself. "It's time, we're ready, guys." The *Medens* shuttle landed unnoticed amidst the commotion on the platform. The sole focus was on the two Mons Administration craft that had now opened their doors. Leena led the *Medens* crew members onto their craft, swapping places with Siggy, who had taken in the view of the broken sun-ray. "Hey, did you see the beam from the mountain hitting Verona Moon? Is that part of the sun-ray?"

"Don't ask, Siggy." The COO gave a signal to take them up.

"Where's Yolanda?" asked Robero with concern, his black-ringed eyes wide and searching. "Where's my boss?"

Olisa took Robero to a back corner of the shuttle, proceeding to console and inform him.

"Take us up now, Leena!" demanded the COO.

Hames could see Leena was struggling to shift the craft. Her hands moved quickly, attempting different methods at motivating the shuttle to lift off. "I can't get it to move. I need to rebalance once more." The new effort at reshaping the outer side of the shuttle was interrupted by a large crack and massive tilt of the landing platform. The tower was crumbling apart.

The areas holding the two Administration shuttles fell away, taking the shuttles with them. The two vessels spiralled down to the ground, along with the people who had been pushing to board them. The ones who avoided this fate ran to the tower in a dangerous attempt to escape through the carnage far below.

The platform immediately beneath the *Medens* shuttle also cracked and Leena took the craft into the air on its own support, hovering above the freshly broken surface.

"Wait!" Emmalyne was pointing at the last unbroken part of the platform. "The middle section there; it's Bas!" Leena swung the shuttle craft around to the central and remaining point of the platform to offer Bas-Mons entry.

Francois pulled him in. "It seems, young man, that we cannot get rid of you." There was no reply from the devastated son of Tar-Mons. His face was blank with shock, his long hair drooping around his face, signalling his defeat.

The COO sat him down next to her. A dusty green-skinned arm wrapped around him. "I'm going to get my people safely up to the *Medens*, okay? Then I can take you over to the other station. The one in the valleys you mentioned." Bas-Mons gave his silent agreement to this propersistion.

The shuttle craft docked with the *Medens* in space, its magnets locking into the larger ship. The remaining members of the crew gazed out the side windows at the fourth moon of New Galapagos. It seemed to be glowing a brighter shade of green every moment or so. It was as if they were standing at the top of a mountain they had all climbed together, proud but mourning the loss of their friends. Hames looked at them all and then out to Verona. It was strange that at this moment they would be so fixated on such a sight. Perhaps it was calming, almost a point of meditation.

Leaving her pilot seat and opening up the back hatch to the *Medens*, Leena was in no doubt about what should be happening next. "We have to turn back to Earth. We've lost too much. It's just too dangerous."

"Let's get the others disembarked. You and I can then take Bas over to the other sun-ray station. After that we can discuss the next steps," said the COO as she urged them off the shuttle.

The crew made their way into the shuttle waiting area. Leena bashed her fist wildly on the portside window at New Galapagos, anger surging in her voice. "Down there that sun-ray station is caved in, destroyed by those crazy villagers. The tower collapsed in on itself and a bizarre power ray is lighting up the sky!"

"Calm down, cousin, we'll discuss it in good time."

"There's nothing to discuss!" shouted Leena.

Hames knew what he wanted. To move on to Mermai, not to return to Earth. Despite all the pain this path had caused, he needed to stay on course. "Once we're in space, what difference does it make if we go one way or the other?"

Tears were in Leena's eyes. "We've lost our friends, Chief! Four of them! Would you risk more of us?"

Olisa tried to comfort her. "We are everything. The Universe will never separate us from Yolanda or Temett." Hames was not surprised by Leena's exasperated look at this consolation.

The COO looked at Hames as she helped Robero through the hatch door. "It's not great, okay? We're here now so let's deal with what's in front of us and take it one step at a time."

Hames had been doing that: taking things one step at a time. It was him who had cut off the villagers' energy source. He reasoned this damage away in the quick thought that the particle conversion drive he had found in the village was due to ignite into catastrophe at any point. In truth, the cube villagers owed him their life. He watched Melini close the shuttle hatch door behind Robero and hoped she would understand this.

Hames' thoughts were interrupted by gasps around him. He looked out the window and saw the fourth moon begin to glow an intensely bright green. The luminescent atmosphere that had previously hugged Verona moon now extended past its aura, reaching out into space and began to part from the moon.

The crew gazed incredulously at Verona and its green light. Melini moved to stand next to Hames. "What's happening?"

"I don't know. I honestly don't know." Hames racked his brains as the green light began to shift upwards to form a shape. It was now completely separate from the moon. The looks of astonishment and wonder slowly turned to horror. Horror at what was happening beside the fourth moon. The ball of energy was large, almost a quarter the size of the moon itself, and at least four times the size of the *Medens*. The aura crackled shades of green and black in its translucent form. Stars shone through its less opaque parts and its shape shivered and morphed in ovals and circles. Then, it began to configure a longer design. The stretched thick body of energy began to move.

Olisa gasped. "Oh, my mind!"

The COO again pressed Hames. "Chief, an answer would be really helpful."

Hames could not speak. They all watched in stunned silence as the energy moved away from the moon and surged like a comet towards New Galapagos. Moving in an oscillating swimming motion, the energy body became progressively faster as it headed towards the planetoid. Their view was obstructed in the shuttle, so they all ran to the repair control room to watch through the monitors brought quickly to life by Emmalyne. The energy being surged over to the Impenetrable Valleys and the partner station for the sun-ray.

Bas-Mons held his hands over his mouth and spoke through his fingers. "No, please, no." The energy being dived into the location of the other sun-ray station, which continued to populate the sky with the circular beam that formerly encased the sun-ray. The energy being covered the whole mountainside in emerald electricity. Explosions and fire erupted. Some of the mountainside faded away. Leena put her arm around a devastated Bas-mons.

Hames felt an answer in not only his head but in his body. "It's the eels. It's like the eels with the egiya readings. You know, the scans we used on the *Medens* to find the Ugor Eels. We found similar readings on the fourth moon when something like that struck down Marak." Hames had the attention of the room and pulled out his med-bar. "When I went to see him just now I scanned him and found a similar reading. He still had egiya inside him, probably from the moon. He..." Hames' last sentence was cut off by another explosion on the far valley as the large mass of egiya energy started to shift from the mountainside.

The being rose into the outer atmosphere above the valleys, almost at the same height of the *Medens* in its orbit, and seemed to pause. Then it turned, once again forming a long oval comet shape. This time it was not heading for New Galapagos. This time its destination was undeniable. It was heading straight for the *Medens*.

Chapter 29

Egiya

MELINI GUERRA RAN TOWARDS THE exit of the repair control room. With her back to her team she beckoned them to join her. "Emmalyne, stay here and monitor where that thing is. The rest of you with me." The crew of the *Medens* sprung to action, all following their leader except for Olisa, who sat pale and drawn, the exertions of their escape from the biotower eventually catching up with her.

Bas-Mons joined them, running after the COO. "You have to take me back down!"

"Not right now, Bas, that thing out there is heading towards us." The COO outpaced her team as they ran through the corridors. Underneath the crew, in the gangways of power lines and connections, sat the main channel of energy and sustenance for the *Medens*. Without power it lay empty and inert. "Hames, Leena, come to the bridge with me. Robero, Siggy and Francois get to the light alignment drive; we have to get it working." The three men stayed with the rest of the crew for a moment before diverting in single file along a right-hand corridor.

"Please, I have to get back down to my people!" implored Bas-mons.

"Afterwards Bas!" The crew ran hard on the path below, their steps rattling out a message of urgency and stress. Melini Guerra pressed her communicator bracelet. "Dao, we're back on board. We need the light alignment drive working again! Get there and help the others."

The voice of the compliant mission technical lead came back to the COO. "Yes. I have finished the repair engine repairs. I will now help the others to fix the light alignment drive."

Leena was petrified but determined. She looked at Bas-Mons, who appeared resigned to his location for the time being. "That thing, it's a monster, Bas. Let's be safe first and then we can figure out what to do next."

Hames' legs were starting to feel their age. He had done more running in the last three hours than he had done for years. He was exhausted and still sore in his head from being knocked out in the cube village. Arriving at the bridge, he was happy to rest at the rear of the half-octagon–shaped room. Leena and the COO took their seats while Bas-Mons paced anxiously.

Acknowledging her presence, the first pilot's seat moved to allow Leena to take her place. As the front panel shifted to accommodate her hands, the small paddle from the back of the seat extended and adjusted itself to align with Leena's Sentir, the large golden circle on her neck. She was connected to the ship and almost instantly the *Medens* shook with a kick to signal movement had commenced.

Leena sought a path for escape from the COO. "What's the plan, Melini? What direction?"

The COO began reviewing screens and data that beamed out in front of her. "At our cruising mode that energy thing is going to catch us pretty soon. We need to get our speed up before we try using the alignment drive, even if to get us a short distance away. For now, head for the first moon – we'll sling shot away."

"Got it." Leena banked the *Medens* away from the planetoid below and out towards the closest moon, a small and haggard-looking rock in the sky.

Olisa finally joined them, the last on the bridge. Still out of breath, she placed a hand on Bas-Mons' shoulder. "What is to become of me?" he asked woefully.

The COO entered calculations as she raised the crew, attempting to bring the light alignment drive back to life. "How's it going back there?"

Robero responded over the communicator, "It's not happening, boss. The LAD needs more alignment testing. If we had a day we could make it work, not today. No chance; it won't hold alignment, not even for a millisecond."

"Damn it!" The COO again sought out more information, bringing a viewer screen up showing the repair control room. "Emmalyne, what's the status of that thing chasing us? It looks bigger from my readings."

"That's because it is getting bigger and faster. It seemed to take something from that station it smashed. I think it wants power. After all, it is an energy being, what else would it eat?" Her frank response sounded more like the old Emmalyne to Hames; it made him smile despite his exhaustion. Olisa, however, seemed to be frowning.

"It's definitely attracted to us." Hames realised their predicament. "It's being drawn to us towards us."

Leena continued to bank the *Medens* towards the first moon, her focus intent on piloting the *Medens* and saving her crew from danger. "What if we shut down completely?"

Hames put his hand through his beard, the broken compass showing to all. "We don't have time for that. We're too powered up with all the systems online to keep the *Medens* cruising. That thing is way too close."

Bas was still reeling yet coming around to the need to protect himself and the ship. "Can we shoot anything at this thing, disperse it?"

The COO continued to concentrate on the data in front of her, a sweat starting to appear on her green-skinned forehead. "We don't have that kind of capacity, Bas."

Hames closed this idea off for good. "Even if we pulsed an energy beam or anything at it, I reckon it would just eat it up like the partner station in the valleys. Let's push on and see if it runs out of steam."

"Not just the first moon," said Leena. "All four moons of New Galapagos provide the cover we need. They nearly line up in right angles; after this we could slingshot through all of them. It might give us some speed to outrun it." Her last statement was more a declaration of hope to the others on the bridge.

The COO shot down this hope. "We need another solution. At cruising mode this green beast will eventually catch us. Even after we have slingshot around all the moons. Running is a temporary option."

Emmalyne came onscreen from the repair control room. "Not temporary. Not if that option is running fast like we used to do. Running faster than people have for decades."

This proposition was initially lost on all of the crew members present, bar Hames. He understood what Emmalyne was suggesting, and he would not be part of it, let alone help it happen.

Melini wiped sweat from the edges of her deep brown hair, its grey tinges showing more noticeably. "What are you saying, Emmalyne? The light alignment drive isn't working."

Emmalyne's flat response betrayed no sense of fear or anxiety. "With Dao's last fixes the repair engine is ready to work."

Olisa voiced the question most of those on the bridge were thinking. "How does repairing damaged space particles help us get away from this being?"

"Urg, it doesn't!" Leena was struggling to keep the *Medens* straight as it hurtled towards the first moon. As they passed into its small gravitational pull she wrenched it hard to bank left. The ship vibrated violently and all the remaining crew gritted their teeth as the *Medens* shuddered and moved on its side. It twisted around the first moon and then lined up for the second.

The crew on the bridge readjusted themselves. Emmalyne's hazel eyes gazed at them from the repair control room. "Yes, for sure, creating engines to repair damaged space took decades. Her head tilted in abject consideration of her plan. "Turning it back the other way around, though, removing some the filtering points, is entirely achievable."

"Emmalyne," the COO called out, "is it even possible what you are suggesting? Wouldn't it take weeks to refit the repair engine?"

"No, not at all. It's mainly reprogramming – the removal of a few sieves and reallocating the conversion plates."

"Is she really suggesting what I think she's suggesting?" Leena pushed the *Medens* hard in a straight line for the next moon.

Hames stayed silent, his fury boiling at Emmalyne. This was the most dangerous option to take and one that would destroy them or lose them forever. Bas was not so silent. "I believe I saw her suggestion earlier today. At the cube village, it was a particle conversion engine that was in use."

"I can't believe you are all considering this!" Hames stood up in anger. "The only reason that village hadn't been destroyed was luck, pure dumb luck. If it had converted any of the damaged space the whole village, maybe the whole planetoid, would have gone up!"

The COO stood up to calm him down. "Now easy, Chief, we know how you feel about this. The village had a particle conversion engine working so it's possible."

"So is jumping out an air-lock, but you wouldn't survive long! Leena, you want us to go back to Earth and stop the mission; that will never happen if we do this!"

But Emmalyne had planted the seeds of strategy in the COO. "Chief, you told me once the space was too damaged around here to travel to Mermai, that you couldn't plot a route. What about somewhere else?"

"There is nowhere else. You can't travel on any of the tentacles in a particle conversion ship; you know this, damn it! That is the whole point of the mission!" Hames' fists were clenched. He knew where this conversation was heading and his young mission colleague in the repair control room took it to the end point.

Emmalyne looked back through the viewing screen. "It's true there are no more mapped routes – they took years to plan out. It doesn't mean we can't use the particle conversion drive to escape. There is clean space past the fourth moon, just not in the direction of Mermai, nowhere near it."

Hames tried one last time to dissuade the COO. "Your family. You will lose all hope of seeing them. If we do this we won't survive, we could end up anywhere. In the dark space, between galaxies, a sun, an asteroid field, a nebula, anywhere but a place where we know where we are!"

The COO shook her head at the chief mission officer with clouded eyes, her voice cracked and broken. "Do you think I like this? We have to live first, Hames."

"You will never get to Mermai, the damaged tentacles will never be repaired. The Organisation will switch off from trying to reach the colonies completely." He looked down at his hands. "I'll never know what happened to her."

Leena growled and in spite of her neck and body being locked into the *Medens* she was able to direct her disdain at Hames. "It's not about getting back to Earth, Hames! It's about looking out for each other. In any case, I don't need to be on Mermai to feel connected to a world where people I know and love are. I feel love now, Chief."

Hames found himself pacing the bridge. "It's different, you don't know what you're missing, who you are missing!"

Olisa joined in the debate against Hames. "We are all connected, Chief. We are always together. It doesn't matter whether you are standing next to someone or sixty light years away from them. We are everything."

The COO turned to her control screen. "Emmalyne, start making the changes. What do you need?"

"I need Dao in the repair control room. I'll take care of the conversion plates and filters in the repair engine room and," she paused, "I need the programming for particle conversion drives so I can alter the interface block. I need Hames."

"No." Hames sat down.

"Fine." Emmalyne left the repair control room on the viewing screen. "I'll have to dig out the programming myself!"

Bas-Mons arose from his seat. "I'll come and help you, Emmalyne, as much as I can." His generosity in the face of all that had happened today made Hames feel remorseful for his recalcitrance. He needed to think of an alternative to this madness. Letting the *Medens* vanish from the known, vanish from the Universe, was not acceptable. Any hope of fixing things, of finding Nisha, would be gone. Searching his mind for answers he found nothing, only frustration.

Leena continued to run the *Medens* hard towards the second moon, her hands obscured by the panel moulded over them as she heaved to the right and turned the *Medens* sharply away from the second moon. For a moment it appeared the egiya being had disappeared, its large shimmering body hidden by the rocky satellite behind them. But as they surged forward so too did the monster, following dead on their trail.

"I can't believe this thing! Is it real?" shouted Leena. "I don't know whether to laugh or cry."

"It's real," grimaced the COO. "Just look at that energy readout – it's like a giant thunderstorm chasing us. It destroyed that station in the valley with ease. We can't let it touch us."

Hames looked over at the viewing screen. Both Emmalyne and Bas were in the repair engine room, working away at the repair engine interface block much like Xavi and Chung had done weeks ago to no avail. They had died and all because of the cobbled-together particle conversion drive in the cube village. After his anger at finding that machine, after causing so much pain and damage in its destruction, how could he accept any role in creating another particle drive engine? It was sacrilege to their memory, and Temett and Yolanda. In any case, what would be the point of living if they could not continue their path? However, another voice in his head was saying, *But what if it could be done?*

"Look!" The COO pointed at the viewer screen as she shouted at Hames. "Look down there! There's your friend, the one you felt so bad for while she adjusted to life on this ship. You stood up for her, spoke up for her. Now look at her, saving everyone's life while you're still stuck in the same place you were forty years ago." The COO was madder than he had seen her before. "I've just as much to lose as you do, Hames."

It was true. But just like the broken compass on his right hand, he had never healed. This was his complete fear and reality in this moment. Nothing in the last forty years had been fixed; only half-attempts. Since the Cataclysm, since he lost Nisha, he had been broken. It was all he knew, loss and an open-ended search for answers. Hames looked at Olisa, who had given Emmalyne Curate-mente. It had seemed to work.

On the viewing screen Bas yanked a filter away from an open section of the repair interface block, just as Leena reset the path of the *Medens* towards the third moon. The shift of the ship and a surge in the gravity well provided an extra element of force that drove the filter flat into Bas-Mons face and once more that day, he was unconscious. It looked as if Emmalyne was unaware of these events and she continued to work away at turning the *Medens* from a ship of repair to a ship of immense speed. She was down there alone. Hames waited for the *Medens* to steady itself, then guardedly he pushed himself out of his seat. Still conflicted and unsure if he could agree to this task, Hames faced the COO. "Let's just wait. Wait to see if we have to do this. There may be other options." He was almost pleading with her.

The COO looked back at him with some form of relief. "I'm open to any option, of course."

"The egiya being may run out of power, it may deviate. Let's give ourselves a chance of staying on track."

The COO focused on the screen showing the energy being chasing them down towards the next moon. "Agreed."

He raced down to the repair engine room and found Emmalyne dragging Bas over to a wall and away from any danger. "I'm sorry I didn't come sooner. I don't agree with this."

She gave him a half-smile. "I think I know that."

His protestations recognised, Hames slipped into his best-known form, that of chief mission officer, in charge of all things related to particle conversion technology. He went straight to a screen on the back of the interface block. "Okay, we can have a go at this, make it safe and make it work. Only if needed."

"If needed," she repeated back to him, with a firm nod.

"Please, a status update."

"We've realigned a lot of the surface conversion plates on the outer part of the ship. Dropped as many of the external sieving points, but still have to pull off some of the internal filters. One of those knocked Bas out."

"Well, I'm no good for that type of work, young woman. You'd better get your tall frame into it. As soon as you see anything wrong, just back away."

"Yes, Chief."

They worked and pulled their way through codes and filters, making all the changes that they could see necessary to turn the repair engine into a particle conversion engine. All the while, Hames felt like he was in his early twenties again, back at the Academy building ships to travel the Universe. It was likely though that this particle conversion ship would condemn them all to death or to the abyss of space.

As the *Medens* hit the apex of its trip around the third moon they both stopped working to take a secured position while the ship was in its unstable state.

Hames stood back from his screen. "It's nearly ready. Dao's confirmed the first part of my code. I just have to put in the final set. It's not perfect but the *Medens* will respond, if the hardware's right."

"It will be," Emmalyne reassured him. "Bridge, how far away is that energy being?"

The COO responded, "It's getting closer and closer. It's like it's getting faster with the moon sling-shots, just like us. It won't take long to reach the *Medens* after we've passed the fourth moon!"

Leena was struggling with the ship. "Argh, the *Medens* has never cruised this fast. I can feel it buckling under the strain. The next manoeuvre is going to test it!"

The *Medens* was indeed shaking as much as Hames had ever felt it. The large glowing interface block shook with tremors as the *Medens* was pushed to its limits. The alarm began to ring out its long warning tone as the highlight bars around the corridors flashed red and orange. It took Hames back to the first accident he and Emmalyne had experienced in the repair control engine room. Working to save the ship while warnings blared was hauntingly familiar. The first time around his goal was to save the mission so they could continue on to Mermai. Now they were making every effort to destroy this dream.

"Chief, I've started the main engine spinning up in anticipation of departure. It's just the coding and the plate alignment to go."

The *Medens* approached the final moon and the fourth banking tack commenced. Hames and Emmalyne stepped back and braced themselves against the wall, holding on for what seemed like hours and not minutes. As the *Medens* shot out on its last path of acceleration it was travelling faster than its cruising mode was designed for. The ship was screaming for release from its metal casing.

The COO hailed them. "It's now or never, team. That thing is gaining on us; about two minutes at best."

Emmalyne joined Hames at the end of the interface block. It glowed ready to accept the final codes and ignite the repair engine. It was prepared to convert space and step light years in seconds. "Hames, you have to do it. Put in the code." Emmalyne's calm demeanour was dissipating, the stress obvious in her hazel eyes as they implored him to send the *Medens* away.

He hesitated. "The COO. Melini said we'd wait until necessary. Has it passed the fourth moon?"

Leena called back over the intercom. "It came around it without blinking, if it could blink. Hames, you must look after us."

The chief mission officer closed his eyes and grimaced. "I can't do it, I can't lose her!"

Emmalyne's voice was rising but she was still in control. "Chief, you lost her years ago."

For all his intellect and experience there was no complex scientific argument to counter the proposition in front of him. "We could destroy all the repair work we have done up to here, all gone in an instant! It's reckless, this one action could destroy the tentacle right back to Earth and god knows what!"

Emmalyne turned her focus completely on Hames, grabbing his arm. "It's the only choice!"

"I won't do it. I won't. It's reckless, we could all die. We could be lost; I could be lost!"

Emmalyne spun around in frustration. "You have to do it, Chief, or we'll all suffer!"

Hames shook his head. "I can't be lost! not like her. How can we find each other if we're both gone?"

The alarms grew louder as the ship rattled more and more, signalling if not the capture of the *Medens* by the egiya being then an alternate fate of being pulled apart by its own forces. The agitated voice of the COO interrupted the cacophony. "We're in free space now! Away from any of the particle debris. Tell us when we can get out of here. This thing is so damn close!"

"We're nearly ready. Soon!" shouted back Emmalyne.

Hames moved back from the data panel, refusing to give away his path in life. "Why do we have to do this? You haven't lived with pain like I have. You just switched it off when it got too much for you!" There it was. An accusation of loss unaccepted, of pain deferred. Grief and loss had been his life. Why should others avoid this reality?

Emmalyne stopped gesticulating and then ran to the interface block to hit a button that closed down the communications to the bridge. Despite the alarms and loud cracks of the *Medens*, Hames could see her shoulders relax, almost feel her breathing out with relief. "I never did it." The tall young woman clung to the interface block as the ship swayed.

"What do you mean?" he shouted back as he moved closer.

"Olisa tried, but I couldn't let her in. I couldn't forget about all the pain of the last few weeks. It hurt too much. The emotions, the ups and downs, the intense time in the LAD turret room. Losing Xavi and Chung. I didn't want them softened. I realised I wanted to feel them, so my head wouldn't let her in."

"You, you seem so strong! You were so anxious before."

"I just needed to talk to someone. I still worry and think about it all. I just kept telling myself I couldn't fix it, that it wasn't for me to fix. Anytime the stress started to make me feel incapable I just did something I was good at, like calculations or fixing the exit at the cube village. The more things I got control over the better I felt about everything."

"What about Marak, he showed you Xavi?"

"He showed me what I needed, to confront what was bothering me. My fears of losing friends, of losing control." The ship shook once more as the alarm sang a final note.

Hames gave her a big hug as if Emmalyne were a long-lost daughter. She pulled back and her face was no longer stern. "Olisa said it was my fault she couldn't get in. She said everyone on the ship would think I was broken somehow."

Hames shouted back at her, "You're not broken! You're perfect!"

Emmalyne grinned back at him. A big smile that lit up her whole face. He could see her, all of her. His mind was still working despite the chaos. "Why didn't you tell me?"

"Olisa has been constantly watching me, standing next to me all the time. You've seen that. I didn't know what to do. Then with the incident at the cube village there wasn't time. Now we have to get out of here." She placed a hand on his shoulder. "The Cataclysm and Nisha, it's okay for that journey to be over. We need you now, not Nisha."

In a moment that could barely be measured in human terms, a switch inside Hames flicked and he moved forward, emotionally and within his soul. Emmalyne was right. As dangerous a plan as this was, there were no other options. Giving him comfort that he held a future was the image he'd received from Marak Chakla of the shrinking silver ball; deep down it told him his journey may not be over. As for the particle drive, he reconciled its use with the realisation that if they were ever found he would rather be alive and in cryo than crushed dead by a giant energy being. He also figured that no damage would filter back to Earth or further along the Mermai Tentacle. Driving the *Medens* away from the fourth moon had meant they would be well clear of any connection to the tentacle and in clean space. Whatever happened, he was ready to help his friends.

Hames entered the last code and then checked the image of the egiya being that was only hundreds of metres away. "I'm done, it's ready. It will only work for about eight seconds. That's enough, let me tell you. We'll be light years away.

Emmalyne switched the communications with the bridge back online.

The COO was furious. "What have you two been doing?

Emmalyne made a quick apology. "Sorry. Hames has finished the work. We're ready, Leena. You have the controls."

The voice of the first pilot was strained. "Hold on tight, everyone. On my mark, now!"

Hames looked to Emmalyne as they held onto the interface block. "Thank you."

The grey panels of the interface block appeared to glisten and the world around them slowed to nothing as if they were caught for eternity in that split second. Hames' eyes concentrated on the egiya being that was directly behind the *Medens* and soon to end them. As the newly constructed particle conversion engine came online it ignited both the space behind them and the being that had so aggressively pursued them. The egiya being turned from brilliant glowing green to a ball of darkened flame. Hames was certain he heard a howl in his mind. Then it happened. The ship shifted worlds, shifted stars and traversed the constellations like a god running through its garden in the night. Hames felt its familiar pull, although the roughness of the makeshift engine was felt too. He blacked out once the ship fell back into reality.

Electricity filled the *Medens'* immediate trail and then dispersed into black. Behind the vessel came waves of obscured distortion that dissipated more and more as the ship began to drift aimlessly in the aftermath of its catapulting from the dangers of the egiya being.

Half-awake, Hames felt the statuesque Emmalyne Biggs helping him walk, his arm around her back. Both of them staggered down to the repair control room. The *Medens'* lights flickered on and off. Beneath their steps the main conduit filled itself to pulse a new but sporadic beat of power. Hames was lighter and happier than he had been in long time. Emmalyne eased him onto one of the tables under the observation dome of the control room. As he lay there he could hear Dao and then the others. The COO, Leena, Bas and Francois, all his friends. They were safe and he was content. It was a feeling of relief unlike he had ever felt before.

Around him he could tell the *Medens'* lights were failing. So too were the screens and panels showing the progress of the repair mission to Mermai. He opened his eyes just as the room went dark. There in front of him and the crew was the unmistakable image of the light years' wide Orion Nebula. Somewhere in front of this was the disconnected colony of Sol-Concio. After all that had happened, they were not lost, just a lifetime away from their original journey and mission. The path of the *Medens* had shifted in an inexplicable change of direction yet it was still pointing towards a lost community of Earth.

In the room half-lit by the immense nebula ahead of them, Hames could see Emmalyne, safe and watching over him. They had survived what he believed was the unimaginable. He had survived the loss of Nisha and the letting go of her trail. At least he was alive and present. Emmalyne had shown him this was possible. Her return smile to him was one of satisfaction. This was the Emmalyne he always knew was there. Not a better Emmalyne nor a different person, just simply her.

Hames laughed to himself. He thought of the promise of Universal Direction and that there was a purpose for everything and a connection for all. Now, right in front of them, there was no path and no connection he could see. Despite this comparison he felt clarity and composure as they all marvelled at their destination. There was a strength in the crew and a belief that while uncertainty lay in front of them, the Universe held the promise of life and more to come.

Epilogue

ONCE MORE THE SHIP TUMBLED over and over. It spiralled downwards through the surge of damaged particles that flowed all around them. Bodies smashed against walls of metal as the ship spun in a wave of light breaking across a dense black sea.

They had not exploded; something else connected to the *Python* had been detonated. Nisha looked up from the floor to the systems above her. They were intact although a massive destruction was flowing behind them like a streak of lighting. She reviewed the screen of pathways showing the tentacles of travel from Earth and its colonies. They were disappearing.

Behind the *Python*, the trail from Earth was illuminated in a blue fireball, crackling back as far as she could see.

Nisha had learned much from Hames and she could see the basic structure of the space was changing. She clung to her mermaid pendant realising the *Python* was travelling far away from her home and far away from him. The bridges to their islands in space had been irrevocably broken.

About the Author

I'm a Tasmanian based writer with a passion for science fiction. I'm fascinated by the question: "what comes next?"

Born and raised in Hobart, I grew up on Star Trek, Star Wars and loads of comics. My interest in science and speculative fiction then took off with early readings such as Fahrenheit 451, Brave New World and Timothy Zahn's Thrawn Trilogy. Recent influences come in the form of Arthur C Clark, Daniel Keys Moran and graphic novels written by Warren Ellis.

For the past sixteen years I've worked across the energy sector in roles based in London, Melbourne and Hobart. Throughout this time I've seen the impact energy can have on people's lives and their broader communities. This experience has greatly influenced my first novel, *Universal Direction.*

Development on this work started in early 2013 and its consideration of how different forms of light and energy may shape our future consciousness was enhanced through the experience of exhibitions by the Museum of Old and New Art (MONA), particularly Spectra by Ryoji Ikeda and Zee by Kurt Hentschlager.

I'm currently working on the next two novels in the Universal Direction trilogy, Book 2, *Reflection of a Dark Moon* and Book 3, *Mind of the Sun.*